ELECTRICAL MEASUREMENTS
Fundamentals • Concepts • Applications

ELECTRICAL
MEASUREMENTS
Fundamentals • Concepts • Applications

Martin U. Reissland

JOHN WILEY & SONS
New York Chichester Brisbane Toronto Singapore

First Published in 1989
WILEY EASTERN LIMITED
4835/24 Ansari Road, Daryaganj
New Delhi 110 002, India

Distributors:

Australia and New Zealand:
JACARANDA-WILEY LTD.
GPO Box 859, Brisbane, Queensland 4001, Australia

Canada:
JOHN WILEY & SONS CANADA LIMITED
22 Worcester Road, Rexdale, Ontario, Canada

Europe and Africa:
JOHN WILEY & SONS LIMITED
Baffins Lane, Chichester, West Sussex, England

South East Asia:
JOHN WILEY & SONS, INC.
05-04, Block B, Union Industrial Building
37 Jalan Pemimpin, Singapore 2057

Africa and South Asia
WILEY EXPORTS LIMITED
4835/24, Ansari Road, Daryaganj
New Delhi 110 002, India

North and South America and rest of the world
JOHN WILEY & SONS INC.
605 Third Avenue, New York, NY 10158, USA

Library of Congress Cataloging in Publication Data

ISBN   0-470-21357-4 John Wiley & Sons, Inc.
ISBN   81-224-0072-8 Wiley Eastern Limited

Printed in India at Prabhat Press, Meerut.

Dmcc

To my Students
AND FOR
FLORIAN

Preface

The idea of writing this book derived from my lectures which I delivered at the University of Siegen, Germany, the University of Dar es Salaam, Tanzania, and the Technical College of Cologne, Germany. Naturally the syllabi of the subject of Electrical Engineering Measurements are more or less the same. This provided the chance for me to implement my experiences which I collected with my students in various places. Without their excellent performance during this course of measurements and the fruitful interactions between them and myself this book would have become much poorer. But now a considerable standard is reached and I have learned that the students can catch up with it. At the beginning of my efforts I thought that too many details might burden the subject unnecessarily. But, after all, I am convinced that technics generally is composed of innumeral details and only by grasping them and their consequences the student will be enabled to collect the necessary knowledge that will make him a successful engineer. The details are quite often of practical nature. Studying them means to take over the experience of others. This way a sound access to the science and art of measurements are available, They cover many fields other than electrical ones, too, but modern technology makes use preferably of electrical methods.

This approach of presentation tries to satisfy the needs of second and third year undergraduate students. Except for some basic knowledge in Mathematics, Experimental Physics, and Fundamentals of Electrotechnics no other references are needed. The derivations usually base on accepted standard knowledge and are further developed by employing approaches which allow an easy access to the actual problem. The discussions of results which normally aim at applications serve the same purpose. Occasionally sample calculations or the presentation of experiments are heading for that goal, too. A numerical, quantitative approach gives a sense of the actual importance of different terms. Reasons like that may allow justified approximations for breaking down a problem into manageable portions without deviating too far from the rated accuracy. If, for instance, a certain disturbance proves to be negligible why should it burden further approaches? — Problems derived from practice are integrated parts within the sequence of presentation. They provide an additional quantified access to the subject matter. This approach is of engineering nature rather than to present separate tutorial chapters.

The design of measuring instruments aims at a certain precision at a reasonable price. In order to match the economic needs the engineer should know the boundaries which limit the application of a certain principle in practice. This is the reason for describing the causes of different errors. Most of these efforts are undertaken for the benefit of the user of instruments but occasionally further detailed error causations are cited to sharpen the eye even for those who want to take part in the proceedings of the state of the art in measurements.

The organization of the book is made up like this to avoid idle repetitions and indications to chapters ahead. These efforts result in a new systematic of presentation. The approach proves to be applicable. It allows a high density of information without sacrificing the easy access to it. This way the level of presentation gets gradually more and more demanding finally satisfying the needs of BSc students to make them professionally fit for measurements.

According to the state of the art analog and digital instruments are equally important. Quite often they are combined in measurement apparatus. So they should be given equal space. The practical background which is carefully underlaid throughout is paid credit to by combining both techniques. Even sophisticated equipment may be made up including sensors for non-electric quantities. Their output voltages or currents may be transformed, transferred, or otherwise be subjected to certain operations. That means at the same time to design or to select special transducers or to place them properly into a measurement system. To meet the challenge which derives from practice is a major goal for the elaborated methodology of the book which also tries to satisfy common academic needs of other fields within the area of technical sciences.

The art of measurements deals with countless disciplines being in need of measurements. Each of them makes normally use of own specialized instruments. This fact may lure a course of measurements into an overloaded presentation with too many details this way loosing the red thread. Or the approach tends to be far reaching so that students cannot find own conclusions. But confining to the concepts by showing the common action principles and basing their presentation on the available fundamentals points the way to quite a level at the end of that course. Students are ready then to find their own way into any specialized application of specific other fields being in need of measurements.

The idea was to stay away from presenting too many instruments. This way enough space was available for derivations and the full description of action principles and their use in didactically chosen examples of applications.

The presentation commences with the fundamental concepts of measurements and an elaborate chapter of measuring errors pointing out their causes and how to minimize them in practice. After that the manifold variety of ordinary instruments is dealt with showing their action principles, deriving their 'scale equations' and describing their technical features and applications and their limits. The most versatile instrument in measurements is certainly the oscilloscope. At the same time it is quite a complex apparatus. In order to make full use of its application range its components should have been understood comprehensively. This is the objective of the related chapter. A strong demand derives from practice to cope with disturbances. Shielding and skilfull grounding in combination with certain means of amplification and processing are common to eliminate interferences from outside and inside the measuring system. The passages of measurement amplifiers present their basic circuits as voltage or current measurement devices providing at the same time a voltage or a current output. This way a very wide application range is covered without the need of explaining elaborate electronics.

Analog to digital converters deal quite detailed with the various voltage to time and voltage to frequency converters. The error causes of offset quantities and interfering frequency disturbances as they effect the measurand are pointed out. The chapter of digital instruments shows the various counter types and digital displays. A quite sophisticated measurement set up of an ampere hour meter makes finally use of many other detailed components which were previously dealt with. It is an application example which shows how to arrange complex measurement apparatus employing basic constituents. This meter presents the highest level reached in this book. The detailed set up involves even the complete electronic circuit this way delivering an imagination of latest demands in the art of measurements to the students. The harmonics in distorted signals may be of serious concern. A chapter dealing with the harmonic analysis of periodical signals copes with the problem in a practicable way. Instrument transformers allow to adapt different devices to each other making use of magnetic coupling. Consequently a detailed approach of measuring magnetic quantities follows up, and a few methods of investigating for magnetic material features are presented.

At the end some mechanical quantities and the one of temperature are selected to show possible applications of electrical measurement methods to investigate non-electric quantities. This chapter is in need of all previously described electrical instruments. They are used processing and indicating the output signal of an electromechanical transducer which picks up the non-electric quantity and converts it into an electric one.

Finally I wish to acknowledge the valuable contributions of my Indian friend, Mr. M.Y. Gujarati (M.Sc.), New Zealand. Being a non-native speaker of English my wording of the manuscript needed some improvements here and there which he helpfully provided. His qualified proof-reading and his various organization talents deserve thanks, too.

I am also gratefully indebted to my colleague and friend, Mr. J. Noerrenberg, Head of Electronics and Telecommunication Section at the University of Dar es Salaam, Tanzania, for his invaluable contributions which he calmly provided, this way supporting my efforts considerably.

Germany, Town of Gummersbach MARTIN U. REISSLAND
October, 1988

Contents

Example

0 Preamble

The importance of measurements can easily be noticed by anybody in everyday life. Measurements are the basis for the understanding of all kinds of deals. The trade of goods is entirely based on well-known quantities. The transfer of information on a certain length needs specific knowledge of an agreed standard of length and the ability to count the units of the measured length. An ancient measure of length (depicted in fig. 0-1) was based on the size of the human foot and was made repeatable by queueing up 16 people leaving church after service and measuring the total length of their feet.

Fig. 0-1.   Historical Depiction of Reproducing the Unit
of Length.

Of course the individual sizes of shoes were quite different from one another, and some doubts about the accuracy of such repeated representations did arise. Sensing this, our forefathers introduced 16 feet to define their specific unit of length. With that they inadvertently—effectively yet—introduced statistical means of averaging, thus meeting the accuracy demands of their times, which were not high anyway. It is obvious that higher accuracy may be obtained by referring to exactly defined fixed standards and abandoning 'natural' or human references. To study nature and to describe it scientifically, one needs exact definitions of references as presuppositions for all measurements. Using and applying these tools to obtain knowledge about nature signifies all research that tries to describe our world for the good use of it by mankind. And this underlines the importance of measurements. They form the basis for all development.

Our understanding of the world we live in is derived from the results of measurement. Galileo Galilei proved by measurements that the earth was a ball and the process of day and night could be explained by its revolution. Galileo revolutionized the world by showing that the sun was the centre of our solar system and not the earth, let alone men. Due to the strong Christian beliefs of those days, people believed it to be quite the opposite. God had created the earth and man, so how could His creation not be the centre of the entire world?

Much later, towards the end of the nineteenth century, doubts arose also about our fixed imaginations concerning the absolute relations among all interactions of nature and, finally, early in the twentieth century, Albert Einstein founded his theory of relativity. In fact, once electrons could be accelerated to very high speeds, reaching nearly the velocity of light, measurements proved the truth of Einstein's theory. The effects caused by these high speed electrons led to the revolutionary conclusion that the mass of matter increases considerably with the speed within that fast range and at the same time, the volume decreases. At the speed of light the mass results in infinity and its geometrical dimensions in zero. This was certainly unbelievable and it was concluded that matter could not reach the speed of light.

The famous German scientist Max Planck once wrote: 'In physics exists only that which can be measured. Our knowledge about nature is composed from many small pieces of innumerable measurements. They do not form a disorderly muddle, but physicists compose them to a sensible understanding of cosmography'. It might even be in need of a change, if new measurement results demand this.

Upto the 1920s, the needs of science were satisfied by assuming absolute relations among all natural processes. But as soon as a higher accuracy of measurements in combination with new technologies became available, new measurement results manifested a more accurate understanding of our world enabling man to proceed even into space. The output of these efforts proved to be the origin of enormous proceedings. Measurements bacame the fundamentals of automatic control techniques. This in turn formed the base for modern automation as soon as electronic technology made economic components of sophisticated structures available. Finally, process techniques are in need of all these techniques together as these are based on measurements (see fig. 0–2).

The different branches of science have evolved their own specialized measuring instruments which meet their specific needs, making the subject of measurements quite versatile. To be successful, the measurement engineer

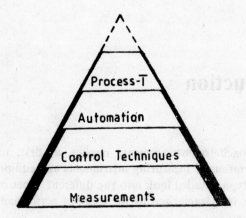

Fig. 0–2. Significance of Measurements for Engineering
Sciences.

should always satisfy two conditions: Firstly, he needs to look into a problem
the way an expert in the field would do and he should feel its needs entirely.
Secondly, he should be able to satisfy the boundary conditions of the
applied measurement methods and always have in mind their application
limits. Successful measurements are usually based on a sound all-round
knowledge and specific details. To fulfil these conditions is no small challenge for the measurements engineer.

1 Introduction

A sensible approach to measurements implies, at first, to deal with the principles of operation of measuring instruments and equipment in a general way. Secondly, a detailed look into the different types of actual instruments. This step is characterized by the study of performance, features and applications.

The operation of a measuring system can be described in terms of functional elements blocks (fig. 1–1). Their output signals have fixed relations with their input signals. A useful function that describes this dependence is the transfer characteristic of each block which characterizes the static and dynamic performance.

Ex.   Moving Coil Ammeter

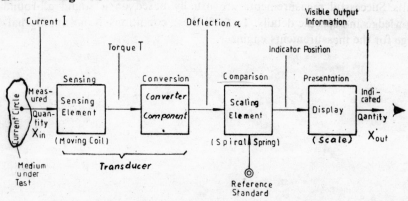

Fig. 1-1   Functional Elements of **an Instrument** (Block-Diagram)

The sensing element receives energy from the medium under test. So it loads the measurand and produces a defined output from the measured quantity. An ammeter (Fig. 1–2) may serve as an example. A current I produces a torque T in a movable coil. The measurand is effected by the loading. Usually, appropriate technical means are employed to minimize this effect.

The converter of the transducer converts one physical quantity into another. For the ammeter the torque T causes a movement α of the moving coil. The new variable α contains the original information (current) in a defined correlation.

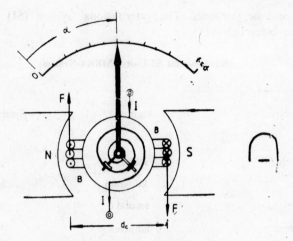

Fig. 1–2. Moving Coil Ammeter.

The scaling element allows comparison with the standard. For the chosen example, the movement is effected against the counter-torque of a spiral spring. It limits the movement of the pointer when the clockwise and the anti-clockwise torques are equal. The equilibrium condition needs to be fulfilled for the final indication angle α of the pointer. The standard is represented by the torque of the spiral spring. By calibrating the scale, the comparison with the current standard is effected.

The display enables the information to be conveyed to the observer. The visual sense is most suitable for an objective scaling and because of its good sensitivity, a high resolution of the reading may be obtained.

1.1 FUNDAMENTAL CONCEPTS AND DEFINITIONS

1.1.1 Nature of Measurements

The event of measuring a quantity is characterized by sensing and representing it, and giving it a certain value by means of a standard. Quantity X is represented by the value of x which is a multiple of the standard $S : X = xS$, i.e. $I = 3\ A$. That means : X is obtained by comparison with the standard S and by counting how often (x) the quantity X contains the standard S. To verify the equation of $X = xS$, the standard S needs to be fixed. This is usually done by convention.

Everyone is familiar with quantities like length, weight, time, etc. Other quantities are in need of special definitions because we cannot directly sense them, such as current, power, etc. Furthermore, quantities such as 'comfortableness' or 'intelligence' are not generally accepted as defined and are therefore not measurable in the sense of the equation for X.

1.1.2 Standards, Units, Symbols

The General Conference on Weights and Measures in 1960 adopted seven units for specific purposes. The international system (SI) of units was accepted as listed below:

Fundamental SI-Units (MKSA-System)

Quantity	Unit	Symbol
Length	metre	m
Mass	kilogram	kg
Time	second	s
Current	ampere	A
Temperature	kelvin	K
Luminous intensity	candela	cd
Amount of substance	mole	mo

Derived (Composed) Units

All other units are defined in terms of fundamental units, for instance :

Pressure $\qquad P = \dfrac{F}{A} \qquad\qquad [P] = 1\,\dfrac{\text{kg m}}{\text{s}^2\,\text{m}^2} = 1\,\dfrac{\text{N}}{\text{m}^2}$

Charge $\qquad Q = I{\cdot}t \qquad\qquad\quad [Q] = 1\,\text{As} = 1\,\text{Cb}$

Work $\qquad W_{el} = U{\cdot}I{\cdot}t \qquad\quad [W_{el}] = 1\,\text{V}{\cdot}\text{A}{\cdot}\text{s} = 1\,\text{J}$

$\qquad\qquad\quad W_{mech} = F{\cdot}s \qquad\quad [W_{mech}] = 1\,\text{Nm} = 1\,\text{VAs}$

There is, of course, no difference, in principle, between W_{el} and W_{mech} for their effects are quite the same, but they may be differently obtained, employing either electrical or mechanical means. As 1 Nm is equivalent to 1 VAs, unit of 1 V, for instance, may be expressed by means of fundamental SI units:

$$1\,\text{V} = 1\,\frac{\text{Nm}}{\text{As}} = 1\,\frac{\text{kg m}}{\text{s}^2}\,\text{m}\,\frac{1}{\text{As}} = 1\,\frac{\text{kg m}^2}{\text{A s}^3}.$$

Due to the lack of a certain immediate notion, it is impracticable to use such composed units. It is better to use 1 V (or F or H etc.) and to relate practical experience with it. That is to investigate what effect 1 V could cause under certain circumstances in a special system. In this way, a fixed imagination of 1 V is formed, which allows the direct practical use.

In practice, we have to deal with a wide range of distances, time, etc., so it is convenient to adopt a number of standard prefixes like

T, G, M, k, (h), (da), (d), c, m, n, p, f, a.

Their meanings are known from the fundamentals of electrotechnics. The prefixes in brackets are in use only in a few countries. Two examples show the use:

$$10^6\,\Omega = 1\,\text{M}\,\Omega \qquad \text{or} \qquad 10^{-3}\,\text{s} = 1\,\text{ms}.$$

Ex. Calculation Procedure for the Use of Quantities

The calculation of an unknown quantity is usually effected using a known relation to other measurable quantities. For example, the power consumed by a heater may be calculated from its known resistance R and the measured current I. Let $R = 10\,\Omega$ and $I = 20\,\text{A}$. So,

$$P = R \cdot I^2 = 10 \cdot \Omega \times 400\,\text{A}^2 = (10\,\text{V/A})\,400\,\text{A}^2 = 4000\,\text{W}.$$

The quantity of the power P consists of its unit (1 W) and a number (4000) carrying the information how often the quantity contains the unit.

Using the units throughout all equations, we are able to control each new step. The units provide a highly appreciated feedback about the validity of an equation. Due to its practical use this procedure is superior to any other method. At the same time, there would be no objection from the mathematical viewpoint.

1.1.3 Direct and Indirect Measurement Methods

Deflection Instruments

Direct measurement methods are characterized by directly obtained measurement values, employing a direct comparison between the measurand and the standard. An example of this may be the weighing of a certain mass, using spring scales (Fig. 1.1.3-1, presenting a simple deflection instrument).

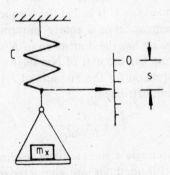

Fig. 1.1.3-1. Spring Scales.

Since the indicated value s represents directly the measured value, the scale of the instrument should be calibrated by a standard of the same nature as the measured object. In this way a comparison can be achieved.

For the final position of the pointer, the forces effecting the deflection s are equal but opposite to each other, giving equilibrium.

$$m_x \cdot g = c \cdot s$$

$$m_x = \frac{c}{g} \cdot s$$

This scale equation shows the that displacement s of the pointer (deflection) from its zero position reflects directly the mass m_x to be measured.

The moving coil instrument may serve as another example (Fig. 1-2). When measuring current I, a comparison of torques takes place. The comparison of the measuring quantity with the standard needs a comparison mechanism that deflects when the measurand is applied. The standard is represented by the torque of the restoring spiral spring. By calibrating the instrument, the comparison with the current standard is realized. The equilibrium of the electrically generated torque T_{el} (clockwise) and the mechanically acting counter-torque T_{mech} (anti-clockwise) causes the deflection angle α of the pointer, giving the actual reading I.

$$T_{el} = T_{mech}$$

$$2. \underbrace{(B \; Il)}_{F} \cdot \frac{d_c}{2} = C \cdot \alpha$$

The electrical torque T_{el} is the product of the force F due to the current I to be measured and the lever length $d_c/2$. But this product needs to be considered twice because of the presence of two air gaps, one to the left and the other to the right of the inner core. The flux density B inside the gaps and the active winding length $l = N \cdot h$ determine the force F. N represents the number of coil windings, and h the height of the coil inside the air gap. The mechanical torque T_{mech} is caused by the deflection α due to the spring rate C. Sometimes C is called the torque direction just to distinguish a linear displacement from a rotary movement. (The spring rate C of a linearly acting spring has the dimension of N/m. But the torque direction C of a spiral spring has the unit of Nm/degree).

Solving the last equation for the measurand I leads to the scale equation of the moving coil instrument.

$$I = \frac{C}{BNhd_c} \cdot \alpha$$

Thus the deflection angle α gives a direct reading of the current.

Indirect measurement methods are characterized by the fact that the quantity to be measured is obtained from physical quantities of different kinds. The measured quantity is calculated from known relations. The realization of derived standards, for instance, is effected by means of indirect methods. The diaphragm pressure gauge of Fig. 1.1.3-2, which is a deflection instrument, may serve as an example of an indirect measurement instrument.

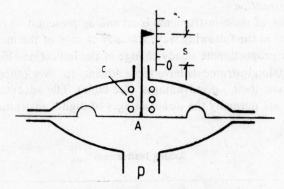

Fig. 1.1.3-2. Diaphragm Pressure Gauge.

Again the opposing forces produce equilibrium at the indicated deflection. 1239.

$$p \cdot A = c \cdot s$$

$$p = \frac{c}{A} \cdot s.$$

The pressure p is transformed into a displacement using a diaphragm of the area A to which p is applied. The displacement of the diaphragm centre is equal to the deflection s of the pointer which can be read as the deviation from its zero position.

These instruments—the spring scales, the moving coil instrument, and the diaphragm pressure guage—produce a deflection of the pointer. That is why they are called deflection instruments. They usually load the measurand, extracting energy from it to cause the pointer movement. In the case of the spring scales and the diaphragm pressure gauge, this is not of practical importance. Once the final indication is reached, all energy conversion ends. But, for the moving coil, a steady current is needed to keep the pointer in its position.

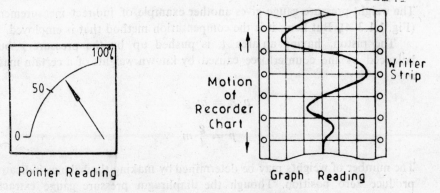

Pointer Reading   Graph Reading

Fig. 1.1.3-3. Analog Displays.

Analog Presentation

The deflection of these instruments is an analog presentation of the measurand, signified in the following way: a steady change of the measured quantity causes a proportionate steady change of the indication (Fig. 1.1.3-3).

While analog instruments have some advantages over other instruments, they also have their   disadvantages (see table). The advantages of analog instruments are normally the disadvantages of digital instruments (see table on p. 13).

Analog Instruments

Advantages	Disadvantages
Easy supervision in control rooms or cockpits	Reading errors (uncertainties).
A curve gives a good view of the past.	Processing and transfer of analog information with limited accuracy only.
Continuous operation.	The deflection is effected by energy which is extracted from the measured medium, thus loading the measurand.
Fast processing and transfer of analog information.	Linearity not easily achieved.

Compensation Instruments

The weight pressure gauge gives another example of indirect measurement (Fig. 1.1.3-4). But here, it is the compensation method that is employed.

The piston, having an area A, is pushed up by the pressure p but balanced by the counterforce caused by known weights of a certain mass m.

$$p \cdot A = m \cdot g$$

$$p = \frac{g}{A} \cdot m$$

The number of weights may be determined by making the balance indicator produce zero position. Though the diaphragm pressure gauge extracts energy from the measurand to cause the deflection, no energy extraction

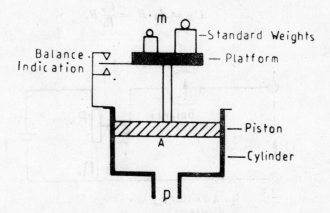

Fig. 1.1.3–4.   Weight Pressure Gauge.

can be noticed after balancing for there is null deflection only. No loading takes place. The compensation of the forces is effected without motion at the balance point.

The balance scales of Fig. 1.1.3–5 make use of quite the same principle.

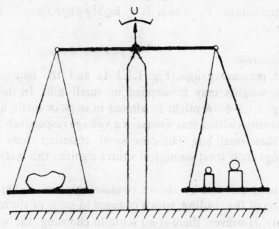

Fig. 1.1.3–5.   Balance Scales.

With the previous examples, the balance scales become quite self-explanatory. The procedure of balancing the scales with the help of weights on the right side works in much the same way as the following one.

Fig. 1.1.3–6 shows a voltage compensator. It serves as an example for an electrical compensation instrument.

The null instrument indicates balance if the voltage drops across the part R of the standard resistor R_s has been made equal to the voltage U_x to be measured:

$$U_X = I_A \cdot R = \frac{U_s}{R_s} R.$$

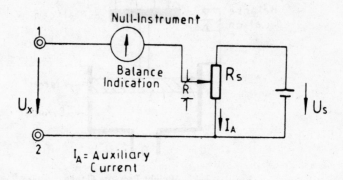

Fig. 1.1.3–6.   Voltage Compensator.

In that case no input current is available as the potential difference across the instrument is zero, resulting in null deflection. This means that the input resistance of the voltage compensator is infinite. So there is no loading of the measurand U_x which is a highly appreciated advantage of all compensators.

Digital Presentation

In the weight pressure gauge (Fig. 1.1.3–4) and the balance scales (Fig. 1.1.3–5), the weights may be applied in small units. In the voltage compensator (Fig. 1.1.3–6), R might be altered in steps of small quantities. The unknown quantity (which was a mass or a voltage respectively) is determined by counting these small bits. The process of counting them and summing up (computing) their total numerical values signifies this method as a digital one.

The counting unit can be chosen considerably small. This results in a high resolution of the reading, which changes in steps of the least significant bit (LSB) only. However, there is no sense in choosing the amount of the LSB smaller than the accuracy of the sensor. The overall accuracy of a digital measuring instrument equals the one of an analog device. It is determined by the sensing element which usually produces a corresponding analog output with respect to the input (X_{in}). But the telecommunication and computation of digital measurement signals can be effected without loss of information. The accuracy of the sensor may be maintained throughout the whole data transfer, up to the indicator.

For indication purposes, a digital display is employed which presents the number of counted units within the measured quantity. Thus a steady change of the measurand causes a stepwise change of the displayed characters. Two examples of digital displays are shown in Fig. 1.1.3–7.

1639.

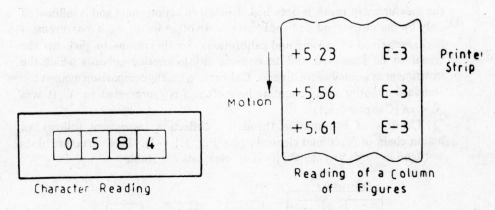

Fig. 1.1.3-7.   Digital Displays.

Digital instruments have their advantages and disadvantages as listed in the table. The advantages of digital instruments are normally the disadvantages of analog instruments (see table on p. 10).

Digital Instruments

Advantages	Disadvantages
No reading errors.	Supervision of changes not easily possible.
High resolution.	Column of figures provides quite difficult access to the notion of the information.
Transfer and processing of digital signals without loss of accuracy.	Discontinuous operation (often just instantaneous samples of the measurand are taken: loss of information).
	Slow processing and transfer of digital information.

1.1.4.   Ideal Block Diagrams

Action Chain in Deflection Instruments
The process of measuring a quantity X_{in} can be presented in an ideal block diagram (Fig. 1–1) independently of the technical realization. It shows that

the measurement result is obtained through an action chain and is influenced only by the measurand and the reference. In other words, each measurement is characterized by sensing and calibrations. Sensing means to pick up the signal to be measured and to convert it into another quantity which the instrument is actually sensitive to. Calibration means comparison against the standard quantity S and counting how often S is represented in X. It was $X = xS$ (Chapter 1.1.1).

The flow of information through a deflection instrument follows an action chain of functional elements (see Fig. 1.1.4–1). This structure does not tend towards any instability if all elements are stable.

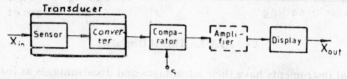

Fig. 1.1.4–1. Action Chain in Deflection Instruments.

The meanings of the following technical terms are self-evident. They are:

— Measuring Quantity	— Measuring Result
— Display	— Measuring Device
— Range of Display	— Measuring System
— Measuring Range	— Measuring Instrument.
— Measuring Value	

To avoid confusion, a brief description may be given as follows: The display is usually realized by the position of a mark (pointer) on a scale. The range of display is the total scale range, whereas the measuring range gives the part of the scale range for which a certain error will not be exceeded. The measuring value is the product of the reading and the unit of the measuring quantity. The measuring result is usually obtained from the measuring value using a certain relation. The measuring device includes all components: sensors, amplifiers, computing elements, display etc. In addition, a measuring system includes parts of the investigated processes that may actuate and guide it. A measuring instrument may be part of a device or a system, or in the simplest case, the device itself.

Action Chain in Null Instruments

The flow of information through a null or compensation instrument is shown in Fig 1.1.4-2. The sensor picks up the input signal X_{in}. It is usually converted to an electrical quantity inside the transducer. Its output is compared with the standard using the null indicator G. It may be observed by a person who cares manually for a zero comparison result (making the known comparison value equal to the unknown quantity). Alternatively the null-amplifier output can be processed by a computer which automatically sets the standard value equal to the unknown quantity. In any case, the part of the standard which is eventually needed to obtain zero at the null indicator is displayed.

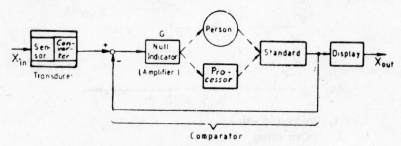

Fig. 1.1.4–2. Action Chain of Null Instruments.

The comparison which is effected at the subtraction point is evident. The difference signal will be small if the null indicator (often a galvanometer G) offers a high sensitivity. Galvanometers are designed for values near zero. Non-linearities for higher deflection are of no importance as the deflection is made zero by the comparator. The balanced device extracts no energy from the process at steady state. The accuracy of the sensor is usually high. To make full use of it, the standard should be of sufficiently high precision, a demand which normally makes compensation instruments expensive.

The action chain of the null method shows the structure of an automatic control loop. It is not necessarily stable even if all components are stable. The compensation process needs time. Endeavours to shorten it may result in an instability of the entire system.

1.2 MEASUREMENT ERRORS

It is important to know the validity of measurement data. Except for the counting of numbers, every measurement involves an error. Whether a certain error appears or not depends on the measurement conditions. The method employed might cause a constant error whereas internal and external disturbances may cause a random change of the reading for constant measurand. The diaphragm pressure gauge of Fig. 1.1.3-2 may serve as an example for that. It is sensitive towards the pressure to be measured, but disturbing quantities may unfortunately cause reading changes like temperature, acceleration, vibration, etc. Different readings exhibit a random scatter, which is due to parallax errors, too. If many small disturbances compensate one another by random combination, the instrument is said to be under statistical control. In reality, this compensation is not perfect and the readings vary randomly around a mean value. In general an error has two constituent components: a systematic consistent one and a random inconsistent one. The results scatter, (see Fig. 1.2-1) for repeated measurements. The mean value of all readings

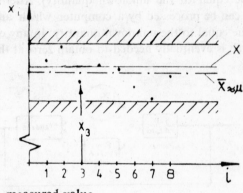

X_i—measured value
$\overline{X}$ —mean value
X —true value

$\left.\begin{array}{c}\text{/////}\\ \overline{}\\ \text{/////}\end{array}\right\}$—Boundary Range: Class of Instrument

Fig. 1.2.–1.   Measurement Result X_i for Each Repeated (*i*) Measurement.

shows a (comparatively) constant deviation from the true value, which is the systematic error. It is determined for its magnitude and its sign. Additionally, there is a random deviation of each result from the mean which changes unpredictably in magnitude and sign. For this random error, only a boundary range may be found. It outlines the limits up to which the random error may occur. It is signified by the class of the instrument which gives the random error in per cent for full-scale deflection.

Whether an error is of systematic or random nature may be determined by repeating the measurements for constant measurand X, i.e. the pressure p.

1.2.1 Systematic Errors
If the measuring conditions deviate from the calibration conditions, the result is a systematic error E_s. It is constant under constant conditions. The error E_s of a certain measurement, may be calculated from the mean value $\overline{X}$ of different measurements, and the true value X, if it is known.

$$E_s = \overline{X} - X$$

As E_s is constant, it is reproducible. In practice, E_s can be determined by employing an instrument of higher accuracy for the determination of X. This means that by comparing the results obtained with the instrument under test and the result of a precision instrument, E_s may be determined.

1.2.2 Random Errors
Random errors cause a change in reading for each repeated measurement. The reading of a pressure gauge may be repeated 10 times. Assume the conditions A and B, for two different instruments A and B. The results are:

$\dfrac{X_{iA}}{\text{bar}}$	8.3	8.5	8.4	8.6	8.5	8.5	8.7	8.4	8.6	8.5
$\dfrac{X_{iB}}{\text{bar}}$	8.60	8.50	8.55	8.50	8.45	8.50	8.40	8.50	8.45	8.55

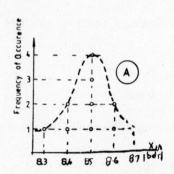 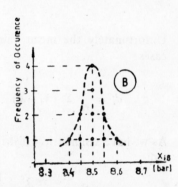

Fig. 1.2.2.-1   Probability Density Functions of Pressure Measurements
under Different Conditions of A and L.

If we arrange the readings from the lowest to the highest and see how many readings fall under each value we get the graphs of Fig. 1.2.2-1, the so-called "Probability Density Functions". The range in which the results scatter is wider for instrument A than for instrument B. This means that B is less sensitive to random influences than A. This fact is also expressed by the number of digits after the decimal point (see table). The frequency of occurrence (probability) that the result X_i lies within a given region is simply the number of times that X_i falls within that region.

If we take many readings and sketch the distribution function, it is found that most random results in practice obey a special function, the Gaussian Distribution Function. There are statistical methods to check whether a set of data is close to the Gaussian Distribution Function or not. But in most cases we simply assume that this distribution is valid. In mass production, the features of resistors or other electronic components or the diameter of balls for bearings have a Gaussian Distribution. However, some random results are of other kinds of distribution. For example, it is most likely that the same number of vehicles are running on tyres that are worn out to the same degree. So the distribution is constant. The so-called Equal Distribution Function is valid. We do not deal with such cases.

1.2.3   Error Calculation

Ex. For repeated measurements under the same conditions (same equipment, same method, etc.) the same systematic error will emerge, but changing random errors can be noticed. The readings of instrument A and B produce the same mean value $\bar{X}$:

$$\overline{X} = \frac{1}{n} \sum_{i=1}^{n} X_i \qquad \begin{cases} \overline{X}_A = 8.5 \text{ bar} \\ \overline{X}_B = 8.5 \text{ bar} \end{cases}$$

But the difference d_i between the mean value $\overline{X}$ and the actual measurement result X_i is greater in most A cases than in B:

$$d_i = X_i - \overline{X}$$

Unfortunately the mean values d_i of these differences equal zero in both cases:

$$\overline{d}_i = \frac{1}{n} \sum_{i=1}^{n} (X_i - \overline{X}) \qquad \begin{cases} \overline{d}_{iA} = 0 \text{ bar} \\ \overline{d}_{iB} = 0 \text{ bar} \end{cases}$$

As we have symmetrical distributions, the following result should be expected:

$$\overline{d}_i = \frac{1}{n} \underbrace{\sum_{i=1}^{n} X_i}_{\overline{X}} - \overline{X} = 0$$

It is self-evident because $\overline{X}$ is the balanced result of positive and negative values.

So neither $\overline{X}$ nor $\overline{d}_i$ can be used to characterize the difference in deviations. But the Gaussian Distribution Function $h(X)$, also called the Probability Density Function, serves our purpose. Fig. 1.2.3-1 shows the graph of $h(X)$.

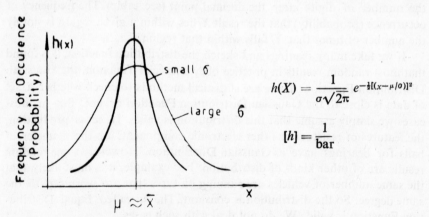

$$h(X) = \frac{1}{\sigma \sqrt{2\pi}} e^{-\frac{1}{2}[(x-\mu/\sigma)]^2}$$

$$[h] = \frac{1}{\text{bar}}$$

Fig. 1.2.3-1. Different Gaussian Distribution Functions.

There are two parameters: μ gives the position of the curve along the X-axis, σ determines its width (the shape). Small σ indicates a high probability that a reading will be close to μ. At the same time, small σ means that there is low probability for large deviations of the actual reading X from the mean value $\bar{X}$. So σ is an appropriate measure to distinguish different random results. (In our case, the dimension of σ is 1 bar). But this method suffers from the need of an infinite number of measurements X_i in order to determine σ:

$$\sigma = \sqrt{\lim_{n \to \infty} \frac{1}{n} \sum_{i=1}^{n} (x_i - \mu)^2}$$

In practice we usually have a limited (n) number of readings. They may be classified such that we get a number (Δn) of readings in each interval with the width of ΔX_i. The plot of the frequency h of occurrence within the respective intervals gives the histogram of the measurement results: step function (Fig. 1.2.3-2). For an infinite number of readings we can choose intervals of a considerably small width, while still having enough readings

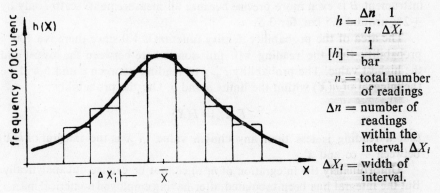

$$h = \frac{\Delta n}{n} \cdot \frac{1}{\Delta X_i}$$

$$[h] = \frac{1}{\text{bar}}$$

$n =$ total number of readings

$\Delta n =$ number of readings within the interval ΔX_i

$\Delta X_i =$ width of interval.

Fig. 1.2.3-2. Histogram of Measurement Results.

in each interval. The step function would become a steady curve, which is the Gaussian Distribution Function (Fig. 1.2.3-1). One can show that in this case ($n \to \infty$) $\mu = \bar{X}$ and $\sigma = S$. S is the sample standard deviation and it serves usually to classify different random measurement results. It is defined as:

$$S = \sqrt{\frac{\sum_{i=1}^{n} (X_i - \bar{X})^2}{n-1}} \qquad \begin{cases} S_A = 0.115 \text{ bar} \\ S_B = 0.058 \text{ bar} \end{cases}$$

Indeed S_A and S_B allow to distinguish the readings from instruments A and B. S gives the mean uncertainty of a single measurement. The mean error $\Delta \bar{X}$ of the mean value $\bar{X}$ and the one ΔS of the standard deviation S may be given as:

$$\Delta \bar{X} = \frac{S}{\sqrt{n}} \quad \text{and} \quad \Delta S = \frac{S}{\sqrt{2(n-1)}}.$$

For a perfect Gaussian Distribution, it can be shown that 68.3% of the readings lie within $\pm\ \sigma$ of μ, 95% within $\pm\ 2\sigma$ and 99.7% within $\pm 3\sigma$. If enough readings are available, $\mu \approx \bar{X}$ and $\sigma \approx S$. In this case, it should be allowed to use, for practical applications, the following approach. Within the range of

$$\left.\begin{aligned} \bar{X} \pm\ \ S: 68.3\% \\ \bar{X} \pm 2S: 95.0\% \\ \bar{X} \pm 3S: 99.7\% \end{aligned}\right\} \quad \text{of the readings will be found.}$$

Ex.   In our case, $\bar{X}$ was 8.5 bar. Taking a deviation of $3\,S$ into consideration, nearly all measurements (99.7%) will be found within the range of $(8.5\pm 3\cdot 0.115)$ bar for instrument A, which means within 8.5 bar $\pm 4\%$. But instrument B is even more precise because all measurements scatter only by $\pm 2\%$ around 8.5 bar for $3\ S$.

The area of the probability density function is 1.0 since there is a 100% probability that the reading will fall somewhere between the lowest and the highest value. The probability $F_{a,\ b}$ of reading between a and b will be the integral of $h(X)$ within the limits a and b. Or: the probability

$$F_{-\infty,x} = F(X)$$

that a reading is less than any chosen value of X is the integral of $h(X)$ between $-\infty$ and X.

Unfortunately the integration of $h(X)$ cannot be carried out analytically. But the integral has been tabulated after having employed numerical means.

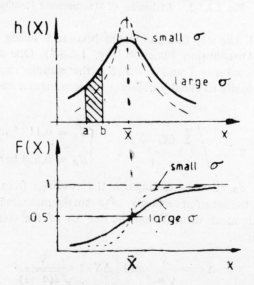

Fig. 1.2.3-3.   Probability Density Function $h(X)$ and Cumulative Distribution Function $F(X)$ of Randomly Scattered Measurement Results.

So the Cumulative Distribution Function (Gaussian Error Integral)

$$F(X) = \int\limits_{-\infty}^{x} h(X)\,dx$$

may be sketched as in Fig. 1.2.3-3.

1.2.4.   Maximum Error in Calculated Results (Error Propagation)

Absolute Error

A result Y may depend on the reading x: $Y = f(x)$. If x is increased by the increment Δx the change ΔY of the function Y may be found using the Taylor series:

$$Y + \Delta Y = f(x + \Delta x) = \underbrace{f(x)}_{Y} + \underbrace{\frac{\partial f(x)}{\partial x}\frac{\Delta x}{1!} + \frac{\partial^2 f(x)}{\partial x^2}\frac{\Delta x^2}{2!} + \cdots}_{\Delta Y}$$

For small increments Δx, terms of higher than first order may be neglected:

$$f(x + \Delta x) = \underbrace{f(x)}_{Y} + \underbrace{\frac{\partial f(x)}{\partial x}\Delta x}_{\Delta Y}$$

Δx may be considered as any uncertainty of a measurement. It causes a related uncertainty ΔY of the result, which is the maximum absolute error resulting from any maximum absolute uncertainty Δx of reading. So far no suggestion was made whether Δx was of random or systematic nature. Both can be considered.

A function z depending on two variables x and y obeys the following relations:

$$z = f(x, y)$$

$$z + \Delta z = z + \frac{\partial z}{\partial x}\Delta x + \frac{\partial z}{\partial y}\Delta y$$

$$\Delta z = \frac{\partial z}{\partial x}\Delta x + \frac{\partial z}{\partial y}\Delta y.$$

Δz is known as the 'total difference' in mathematics. In measurements, it is the maximum possible error. If the signs of Δx and Δy are known, they should be considered (systematic error), if not, the worst possible combination has to be assumed (random error).

Ex. Two examples are given to illustrate the point. Firstly, the power P consumed by a resistor R should be determined from voltage and current measurements U and I. The readings suffer from absolute random uncertainties of ΔU and ΔI at the utmost, due to the class of the instruments. The maximum absolute random error ΔP effecting the power $P = U \cdot I$ may be determined using the last formula Δz which is ΔP:

$$\Delta P = I \cdot \Delta U + U \cdot \Delta I$$

Due to the nature of random errors the signs of ΔU and ΔI are not known. So they may simply be assumed both as positive or both as negative resulting in the worst case each. Consequently, ΔP can be considered as $\pm \Delta P$ only.

Secondly, the resistance R may be determined from the same measurements of U and I as $R = U/I$. The absolute error ΔR follows from the total difference Δz which is equal to ΔR.

$$\Delta R = \frac{1}{I} \Delta U - \frac{U}{I^2} \Delta I$$

The mathematical procedure produces the minus sign. As soon as the signs of ΔU and ΔI are uncertain the worst case of their combinations needs to be assumed to determine the maximum absolute resulting error ΔR. This means that if ΔU is considered as positive, ΔI must be considered as negative. Taking ΔU as negative forces ΔI to be positive. If only random errors are to be taken into account, it is easier to use ΔR as follows, and not to care for the signs of ΔU and ΔI. This pragmatic approach avoids mistakes, but takes into account that the readings U and I are taken from different instruments

$$\Delta R = \frac{1}{I} \Delta U + \frac{U}{I^2} \Delta I$$

Ex. The results for the sum $S = U + V$ and the difference $D = U - V$ should be given, too:

$$S = U + V \qquad\qquad D = U - V$$
$$\Delta S = \Delta U + \Delta V \qquad\qquad \Delta D = \Delta U + \Delta V$$

The plus sign in ΔD was employed for the same reason as for ΔR earlier. Both results show that in sums and differences, the absolute random errors of each term are to be added to find the resulting total absolute random error (in case the readings originate from different instruments).

If several variables u, v, w, x, y determine a result z it is quite unlikely that all random errors cause simultaneously the worst case. Better practical results are produced by the following formula because of random compensation of some error causes:

$$\Delta z = \sqrt{\left(\frac{\partial z}{\partial u} \Delta u\right)^2 + \left(\frac{\partial z}{\partial v} \Delta v\right)^2 + \left(\frac{\partial z}{\partial w} \Delta w\right)^2 + \left(\frac{\partial z}{\partial x} \Delta x\right)^2 + \left(\frac{\partial z}{\partial y} \Delta y\right)^2}$$

Δz may be called the "overall statistical boundary". (It is known as Error Propagation). Its elements Δu, Δv, Δw, Δx and Δy are not considered as absolute maximum limits but rather as statistical bounds.

Relative Error

The relative error is simply the absolute error referred to the resulting value. So the following results ε_S and ε_D for the sum S and the difference D may be obtained.

$$\varepsilon_S = \frac{\Delta S}{S} \qquad\qquad \varepsilon_D = \frac{\Delta D}{D}$$

$$= \frac{\Delta U + \Delta V}{U + V} \qquad\qquad = \frac{\Delta U + \Delta V}{U - V}$$

Replacing $\Delta U = U \cdot \varepsilon_U$ and $\Delta V = V \cdot \varepsilon_V$ produces

$$\varepsilon_S = \frac{U \cdot \varepsilon_U + V \cdot \varepsilon_V}{U + V} \qquad\qquad \varepsilon_D = \frac{U \cdot \varepsilon_U + V \cdot \varepsilon_V}{U - V}$$

For the difference D it should be pointed out that ε_D becomes unacceptably large for small difference D between the measurands U and V. In order to avoid this, high accuracies of the measurements of U and V need be realized, which usually makes such measurement equipment expensive.

Ex. The flow Q of a liquid can be determined from the difference of the pressure p_1 in front of an orifice plate and the one p_2 behind it. Both readings may contain an error of $\varepsilon = 1\%$.

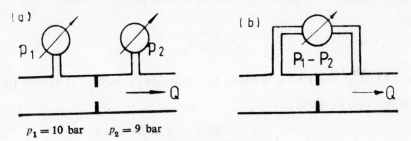

$p_1 = 10$ bar $\qquad p_2 = 9$ bar

Fig. 1.2.4-1. Flow Meter.

Let us calculate the relative error ε_Q contained in Q.

$$Q = c(p_1 - p_2) \qquad \varepsilon_Q = \frac{cp_1\varepsilon + cp_2\varepsilon}{cp_1 - cp_2}$$

$$= cp_1 - cp_2 \qquad = \frac{p_1 + p_2}{p_1 - p_2}\varepsilon = \frac{10 + 9}{10 - 9} \cdot 0.01 = 19\%.$$

Though the single measurements are quite accurate (1%) the result is of poor accuracy (19%) as p_1 and p_2 are close to each other. Instead of separate pressure measurements a transducer should be used which is directly sensitive to the pressure difference $(p_1 - p_2)$ [Fig. 1.2.4-1 (b)].

For the product and the quotient, the previous examples of power and resistance determination from voltage and current measurements should be used again:

$$\varepsilon_P = \frac{\Delta P}{P}$$

$$= \frac{I \Delta U}{U \cdot I} + \frac{U \Delta I}{U \cdot I}$$

$$= \frac{\Delta U}{U} + \frac{\Delta I}{I}$$

$$\varepsilon_P = = \varepsilon_U + \varepsilon_I$$

$$\varepsilon_R = \frac{\Delta R}{R}$$

$$= \frac{\Delta U}{I} \cdot \frac{I}{U} + \frac{U}{I^2} \Delta I \cdot \frac{I}{U}$$

$$= \frac{\Delta U}{U} + \frac{\Delta I}{I}$$

$$\varepsilon_R = \varepsilon_U + \varepsilon_I$$

Both results show that in products and quotients, the relative random errors of each term are to be added to find the total relative random error (in case the readings originate from different instruments).

At last, the relative random error for a function of the following form should be given:

$$f(x, y, z) = K \cdot \frac{x^a \cdot y^{-b}}{z^c}$$

$$\varepsilon_f = \frac{\Delta f}{f} = \pm \left(|a| \frac{\Delta x}{x} + |b| \frac{\Delta y}{y} + |c| \frac{\Delta z}{z} \right)$$

Ex. The special function f, depending on the variables A, B and C serves as an example for calculating the total relative random error $\varepsilon_f = \frac{\Delta f}{f}$:

$$f = 5 \frac{A \cdot B^2}{C^3}$$

As $\frac{d(\ln f)}{df} = 1/f$ it is certainly true that $d(\ln f) = df/f$. Let us form $(\ln f)$ to find $d(\ln f)$ which is ε_f.

$$\ln f = \ln 5 + \ln A + 2 \ln B - 3 \ln C$$

$$\varepsilon_f = \frac{df}{f} = d(\ln f) = 0 + \frac{dA}{A} + 2 \frac{dB}{B} - 3 \frac{dC}{C}$$

$$\varepsilon_f = \frac{\Delta f}{f} = \pm \left(\left| \frac{\Delta A}{A} \right| + 2 \left| \frac{\Delta B}{B} \right| + 3 \left| \frac{\Delta C}{C} \right| \right)$$

Summary of Error Propagation
$X_1(1 \pm \varepsilon_1)$ and $X_2(1 \pm \varepsilon_2)$ may be the variables of $y = f(x_1; x_2)$. The absolute error Δy and the relative error ε_y were obtained as shown in the table on page 25.

Mathematical Operation	Function $Y = f(x_1; x_2)$	Random Error — Absolute ΔY	Random Error — Relative ε_Y
Sum	$Y = (x_1 + x_2)\left(1 \pm \dfrac{x_1\varepsilon_1 + x_2\varepsilon_2}{x_1 + x_2}\right)$	$\pm\,(\Delta x_1 + \Delta x_2)$	$\pm\,\dfrac{x_1\varepsilon_1 + x_2\varepsilon_2}{x_1 + x_2}$
Difference	$Y = (x_1 - x_2)\left(1 \pm \dfrac{x_1\varepsilon_1 + x_2\varepsilon_2}{x_1 - x_2}\right)$	$\pm\,(\Delta x_1 + \Delta x_2)$	$\pm\,\dfrac{x_1\varepsilon_1 + x_2\varepsilon_2}{x_1 - x_2}$
Product	$Y = x_1 \cdot x_2\,[1 \pm (\varepsilon_1 + \varepsilon_2)]$	$\pm\, x_1\, x_2\,(\varepsilon_1 + \varepsilon_2)$	$\pm\,(\varepsilon_1 + \varepsilon_2)$
Quotient	$Y = \dfrac{x_1}{x_2}\,[1 \pm (\varepsilon_1 + \varepsilon_2)]$	$\pm\,\dfrac{x_1}{x_2}(\varepsilon_1 + \varepsilon_2)$	$\pm\,(\varepsilon_1 + \varepsilon_2)$

1.2.5   Static Performance Characteristics of Instruments

Several terms are frequently used to describe the static performance of an instrument. They are defined in short as follows:

Error is the deviation of the measured quantity from its true value. It may be given as absolute or relative error. In the first case it has the unit of the measured quantity; in the second it is the percentage of the quantity measured.

Repeatability is the closeness of a group of repeated measurements under the same conditions. It is given as the percentage of the actual true measurand.

Drift is a gradual output change of an instrument without an input change.

Uncertainty gives the range within which the true value is estimated to lie.

Accuracy relates the actual measured value to the true value implying declared probability limits.

Resolution is the smallest increment of the measurand which still causes a detectable reading.

Hysteresis is the largest plus-minus resolution as deviation from the true value to cause detectable responds for increasing and decreasing measurands.

Threshold gives the smallest input increment from zero to cause a detectable reading.

Zero stability describes the ability of the instrument to return to zero reading after the measurand has returned to zero. The zero setting of good instruments should not be effected by temperature changes, vibrations, etc.

Sensitivity is simply the quotient of the reading change and input change causing it.

Linearity means constant sensitivity throughout the whole measurement range.

2 Moving Coil Instruments

Like all other electrical measurement devices moving coil instruments are used to transform a quantity, for example a voltage, into a form recognizable to a human observer. This is usually realized by a pointer moving over a scale as such indicating directly the value of the voltage, current, power, etc. The deflection is a continuous function of the quantity.

2.1 DESIGN OF MOVING COIL INSTRUMENTS

The moving coil instrument consists of a number of functional elements which as a whole realize its function.

2.1.1 Scale and Pointer

The scale is the most obvious part of an instrument. Besides its main feature to obtain the reading by comparing the position of the pointer with respect to the scale divisions, it carries other important information about the unit of the measured quantity, its nature, whether it is a **DC** or an **AC** measurand, the class, frequency range, internal resistance, etc., (Fig. 2.1.1-1) In chapter 6.2 further details about the symbols in use are given.

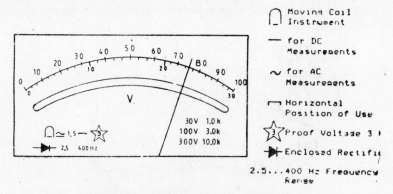

Fig. 2.1.1-1. Scale of Pointer Instruments.

The pointer of an instrument indicates the position of the moving coil, that actually effects the movement. To obtain a fast response to the measurand, the pointer should be light. It may be shaped as a knife-edge or a needle pointer for precision instruments or as a lancet or bar pointer for robust devices (Fig. 2.1.1-2).

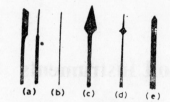

Fig. 2.1.1-2   Types of Pointers
(a) Knife Edge Pointer
(b) Needle Pointer
(c) and (d) Lanzette Pointer
(e) Bar Pointer

2.1.2   Moving Coil, Damper and Restoring Spring

The moving coil consists of a number of windings of copper wire usually wound on a light rectangular frame made from plastics or aluminium. (Voltmeter coils have lots of windings to provide a high internal resistance. Ammeter coils have only a few heavy windings to provide a low internal resistance.) The torque inertia of the moving parts should be low. For this reason DC-instruments have aluminium frames which act as an eddy current damper while moving through the air gap of the magnetic path. For AC-measurands the frame cannot be of a conductive material for it would act as a shorted winding which would extract a lot of energy from the measurand (thus loading it) and eventually overheat. So a plastic frame serves the purpose and the damping is provided by additional elements, such as air or eddy current dampers (Fig. 2.1.2-1 and 2.1.2-2). They unfortunately add an additional torque inertia, which slows down the response of the instrument.

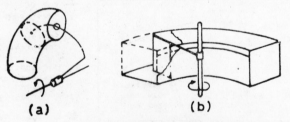

Fig. 2.1.2-1   Air dampers
(a) Piston Type
(b) Blade Type.

In air dampers some kind of a piston or a blade shifts air from one side of the cylinder chamber to the other while moving, thus extracting energy from the movement and changing it. The eddy current damper provides the same result by moving a conductive disc within an air gap of a permanent magnet system. Whatever type of damper is in use it is mounted on the system shaft that carries the moving coil and the pointer as well.

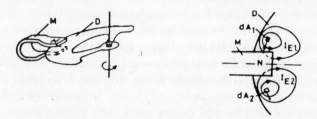

Fig. 2.1.2-2.   Eddy Current Damper.
M = Magnet, D = Eddy Current Disc.
dA = Area Elements,
I_E = Eddy Currents.

The restoring spring is usually wound as a spiral, which provides the counter torque. In many applications there are two springs positioned at each side of the moving coil. They serve as leads, too, for carrying the current to be measured. (The material of the springs is composed of a sophisticated NIVAROX or NIVAFLEX alloy consisting of many different components to provide a constant spring rate over a wide temperature range.)

2.1.3   Bearings and Suspensions
Bearings are the cheapest kind of a suspension of the shaft (Fig. 2.1.3-1). However, other suspensions are quite commonly in use as well.

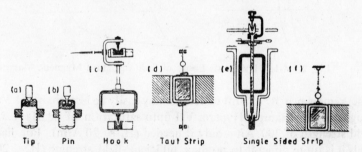

Fig. 2.1.3-1.   Bearing and Suspensions.

The simplest kind of a pivot consists of a pointed shaft tip resting in a cone [Fig. 2.1.3-1 (a)] which usually is made from artificial saphire. The friction of such a bearing is very low but the pressure of the touching area is high, restricting its use to smooth applications. Fig. 2.1.3-1 (b) shows a similar design, but due to the different construction as a shaft it may even tackle mechanical shocks. Fig. 2.3.3-1 (c) depicts a suspension that allows the moving system to be hooked on to the upper tip. During operation the lower tip should not touch the outer ring. The system needs to be strictly horizontal. The system of Fig. 2.1.3-1 (d) is freely suspended between two taut strips. They may provide a considerable small spring rate at the same time being quite robust due to additional mechanical terminals (not depicted) that may take hard shocks from the strips. As there are two strips they usually serve as electrical leads, too. The single sided strip, Figs. 2.1.3-1 (e) and (f), produces the softest spring rate. It is used for most sensitive galvanometers. The QUAMBUSCH configuration (e) prevents damage to the strip due to its upper and lower tip that fixes the system firmly in case of shocks. Normally the taut strips bear the moving coil, and the tips do not touch their counter cones.

2.1.4. Magnetic Path

The indicator deflection of a moving coil instrument is effected by the measured current, passing through the coil. It is subject to a magnetic field usually generated by a permanent magnet. To achieve high induction within the air gap certain measures are taken, regarding materials and construction of the magnetic path.

Magnetic Materials

Magnetic conductive materials differ from each other regarding their different inductions for a certain field strength. Soft materials produce a high induction for low field strength, hard materials usually show somewhat lower induction for considerably higher field strength (Fig. 2.1.4-1).

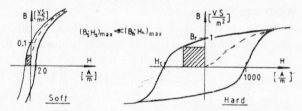

Fig. 2.1.4-1.   Hysteresis Loops of Soft and Hard Magnetic Materials.

The remanence B_r of soft materials may easily be less than 0.1 Vs/m². To magnetize for example Hyperm VII upto saturation of nearly 1.5 Vs/m² a field strength of 300 A/m only is needed ($H_c \approx 20$ A/m). For hard materials it is quite difficult to produce sufficient field strength ($H_c = 200...1500$ A/m) to achieve zero induction after they had been magnetized previously ($B_r = 0.4...1.6$ Vs/m²).

Electro-Magnetic Equivalents

Magnets consist of hard magnetic material. They serve to generate the magnetic flux in the air gap within which the system coil moves. In order to obtain equations for the purpose of dimensioning the magnetic path, it is quite useful to realize the equivalents of the magnetic and the electric field. For the electric field, energy is needed to keep current going whereas it is not needed for the magnetic field to maintain a flux. The equivalent quantities are:

Electric Field		Magnetic Field	
Current	I	Flux	ϕ
Voltage	U	Magn. Motive Force	V_m (Circulation)
Resistance	R	Reluctance	R_m
Current Density	S	Flux Density	B (Induction)
Field Strength	E	Field Strength	H

Usually the load characteristic, (Fig. 2.1.4-2) of the current circuit is depicted as I vs. U within the first quadrant, because current and voltage are understood as quantities of the load R and not of the generator. But for the magnetic path usually the quantities inside the magnet (which is the generator of the flux) are depicted. Consequently the third quadrant is used for the purpose.

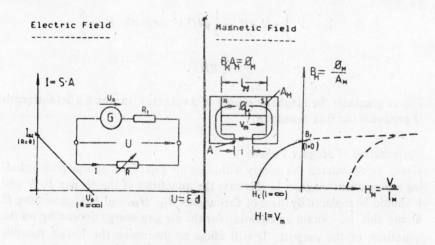

Fig. 2.1.4–2.   **Load Characteristics of an Electric Field Circuit and a Magnetic Field Path.**

Magnetic Energy (inside the volume $V = A \cdot l$ of the air gap)
A current passing a coil produces the magnetic energy $dE = u.i\, dt$. The voltage u across the coil is opposite to the elctromotive force e. So

$$dE = -e \cdot i\, dt$$

$$= N \frac{d\phi}{dt} \cdot i\, dt$$

$$E = \int Ni \, d\phi = \int \underline{Ni} \, A \, dB$$

Energy, whether produced from a current passing through a coil or from a permanent magnet is quite the same, as the magneto motive force

$$V_m = Ni = Hl$$

anyway. The letter l stands for the length of the fieldlines

$$\left(\text{generally } Ni = \oint \mathbf{Hdl} \right).$$

So the last equation can also be written as:

$$E = \int \underline{Hl} \, AdB$$

Using

$$dB = \mu \cdot dH$$

and

$$V = A \cdot l$$

we obtain

$$E = V \int \mu H \, dH = \frac{1}{2} \mu H^2 \, V \quad \text{or with} \quad H = \frac{B}{\mu}$$

$$E = \frac{1}{2} BHV.$$

This is generally the magnetic energy of a volume V in which a field strength H produces the flux density B.

Optimization of Magnet Volume

E may be considered the energy within an air gap of a magnetic path including a permanent magnet. In this case the quantities of the air gap B, H and V should be replaced by those of the magnet B_M, H_M, and V_M, generating E. Doing this we obtain an equation for the air gap energy depending on the quantities of the magnet. It will allow to determine the lowest possible magnet volume $V_M = l_M \cdot A_M$ to obtain a certain air gap energy E.

The magnetic path of Fig. 2.1.4–3 shows that a part of the magnetic flux will be found within the air gap. The stray factor η determines it:

$$\eta \phi_M = \phi$$

$$\eta A_M B_M = AB$$

$$B = \eta \frac{AM}{A} B_M \tag{a}$$

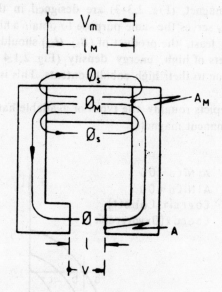

Fig. 2.1.4–3. Magnetic Path.

For the pole pieces an iron of high magnetic conductivity μ (permeability) is used. As is far higher than the one of air the magneto motive force across the magnet V_m will be the same as the one V across the air gap. As

$$\oint H d1 = 0,$$ for there is no current, we obtain

$$H_M l_M - H l = 0$$

$$H = \frac{l_M}{l} H_M \qquad\qquad (b)$$

The equations (a) and (b) are used to replace B and H of the energy equation:

$$E = \frac{1}{2} \eta B_M H_M \underbrace{\frac{\overbrace{A_M \cdot l_M}}{A \cdot l}}_{V} \cdot V$$

$$E = \frac{1}{2} \eta (B_M H_M) \cdot V_M$$

If a large energy E inside the air gap is needed, the geometry of the magnetic path should be chosen so as to facilitate low stray losses. That means η should be near to 1, which is preferably reached by a construction that positions the magnet near to the air gap. Instruments employing a magnetic path with a core magnet, (Fig. 2.3–3) are designed in this way. A large magnet volume V_M serves the same purpose to obtain a high air-gap energy. Last, but not the least, the product of $(B_M \cdot H_M)$ should be high, which is the case for magnets of high energy density (Fig. 2.1.4–4). But they are quite expensive, due to their high cobalt contents. This is why V_M needs to be designed small.

Fig. 2.1.4–4 depicts roughly the types of available hard magnetic materials used for permanent magnets.

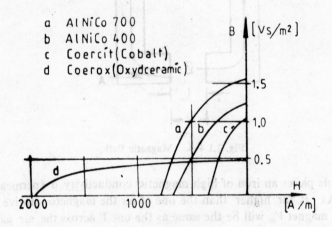

Fig. 2.1.4–4.   Magnetic Properties of Different Materials
Demagnetization Curves.

All materials are mechanically very hard. They can be shaped by grinding only. The oxide types, like d, are far cheaper than the metal types. They also provide a high maximum product of $(B_M \cdot H_M)_{max}$. Additionally they are extremely safe against disturbing fields from outside. Their magnetic state cannot easily be changed. Many other magnetic materials are also available.

Anglo American manufacturers of magnetic materials still use old units for B and H. They are related to the SI-units as follows:

$$\text{For } B: 1G = 10^{-4} \frac{Vs}{m^2} \quad \text{and} \quad \text{for } H: 10e = \frac{250}{\pi} \frac{A}{m}$$

Calculation of Magnet Measures for Certain Air Gap Induction
High energy materials produce a higher $(B_M \cdot H_M)_{max}$ product than others. But more important is a proper dimensioning of the magnet (A_M, l_M) for given air gap dimensions (A, l) and its induction (B) to make $(A_M \cdot H_M)$ a maximum. The procedure is as follows:

A certain magnetic material is chosen. Its magnetizing curve will be known in the second quadrant. From $H_M - B_M$-pairs the products $(B_M \cdot H_M)$ are calculated and sketched as a function of H_M (Fig. 2.1.4–5). Its maximum gives the optimum values of H_{MO} and B_{MO} of the best working point 0.

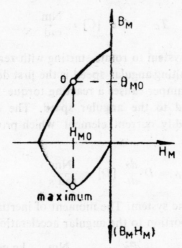

Fig. 2.1.4–5. **Magnet Magnetizing Curve and Its Product Curve** $(B_M \cdot H_M)$ Over H_M.

They allow to calculate A_M and l_M from equations (a) and (b):

$$A_M = \frac{B}{B_{MO}} \cdot \frac{A}{\eta}$$

$$l_M = \frac{H}{H_{MO}} \cdot l$$

Again: **B, A, H,** and **l are the quantities of the air gap.**

2.2 PROPERTIES OF DAMPED MASS–SPRING–SYSTEMS

2.2.1 Step Response

The response of an instrument is the reaction towards the excitation caused by the measurand. The investigation of the time behaviour is quite easy to understand by applying a step input (either switching on from zero to a certain value or switching off) and to study the response of the output. Let us consider a moving coil instrument. Once the current to be measured has been applied it produces an electrical torque T_{el} which moves the coil clockwise. But the pointer which is fixed to the coil cannot respond immediately as three reacting torques delay its movement. The pointer gradually reaches its final position. The interplay of different torques is as follows:

Firstly T_{el} twists the spral springs proportional to the deflection angle α. This produces the restoring control torque T_C. The 'stiffness' C of the spring determines the actual amount of T_C. Sometimes C is named the spring rate. But as we deal with a rotary movement C should be called the torque direction.

$$T_C = C \cdot \alpha \quad [C] = \frac{\text{Nm}}{\text{rad}}$$

Secondly T_{el} causes the system to rotate starting with zero velocity. At any instant of time the resulting angular speed is the first derivative of α with respect to time. So the damper causes a reacting torque T_D that damps the movement directly related to the angular speed. The damper may be a damping blade or an eddy current element which provides the damping constant D.

$$T_D = D \cdot \frac{d\alpha}{dt} \quad [D] = \frac{\text{Nm}}{\text{rad/s}}$$

Thirdly T_{el} accelerates the system. The moment of inertia θ causes a reacting torque in direct proportion to the angular acceleration.

$$T_\theta = \theta \cdot \frac{d^2\alpha}{dt^2} \quad [\theta] = \frac{\text{Nm}}{\text{rad/s}^2} = \frac{\text{kg m}^2}{\text{rad}}$$

Theoretical Analysis of the Step Response
According to Newton's fundamental law the sum of reacting torques equals the exciting electrical torque at any instant of time. The applied current I causes T_{el}.

$$\theta \frac{d^2\alpha}{dt^2} + D \frac{d\alpha}{dt} + C\alpha = T_{el} = K_s \cdot I(t).$$

This differential equation describes the behaviour of the system. As it is of second order we say the instrument is a second order system. Its pointer produces a time dependent deflection $\alpha(t)$. Having switched on a current the pointer reaches its final (static) position after a certain time. All derivatives of α finally become zero. But in between a position transition of the pointer is followed. Anyway the instantaneous change in the value of α is of interest. It is the solution to the above differential equation.

Whether the system is switched on or off is not of substantial importance for its behaviour. The responses $\alpha(t)$ differ quantitatively only. But as it is easier to handle a differential equation which is zero, we will therefore consider the homogeneous solution only. This will be the resulting $\alpha(t)$ from a switched off current:

$$\frac{d^2\alpha}{dt^2} + \frac{D}{\theta} \frac{d\alpha}{dt} + \frac{C}{\theta} \alpha = 0.$$

Two abbreviations prove to be quite useful to change the differential equation into its standard form. They are:

$$\frac{C}{\theta} = \omega_0^2 \text{ and } \frac{D}{\theta} = 2\omega_0 d \quad *$$

*) Proof for the validity of this equation:

$$d = \frac{D}{2\sqrt{\theta C}} \quad \text{relative damping coefficient}$$

$$\rightarrow \frac{D}{\theta} = 2\omega_0 d = 2\sqrt{\frac{C}{\theta}} \cdot \frac{D}{2\sqrt{\theta C}}$$

$$1 = 1.$$

The quotient C over θ proves to be the squared natural circular frequency and d is the relative damping coefficient (in short the damping ratio).

$$\frac{d^2\alpha}{dt^2} + \underbrace{2\omega_0 d}_{a} \frac{d\alpha}{dt} + \omega_0^2 \alpha = 0$$

$$\left(\frac{1}{a} = \tau = \text{time constant of the system} \right)$$

a is the declining constant, $[a] = 1/s$. To solve the differential equation, Euler's hypothetical expression may serve:

$$\alpha = e^{kt} \qquad \frac{d\alpha}{dt} = ke^{kt} \qquad \frac{d^2\alpha}{dt^2} = k^2 e^{kt}$$

Using these statements we obtain the characteristic equation of the system

$$(k^2 + \underbrace{2\omega_0 dk}_{a} + \omega_0^2)\, e^{kt} = 0$$

Only the term in brackets can be zero which is the case for

$$k_{1/2} = - \underbrace{\omega_0 d}_{a} \pm \underbrace{\sqrt{\omega_0^2 d^2 - \omega_0^2}}_{\underbrace{\phantom{\sqrt{\omega_0^2 d^2 - \omega_0^2}}}_{-\omega^2}}^{a^2} = \omega_0(-d \pm \sqrt{d^2 - 1})$$

Three cases for d need to be distinguished:

$$d < 1 \qquad d > 1 \qquad d = 1.$$

In the first case k_1 and k_2 are complex and the solution α of the differential equation will contain exponentially declining circular functions. They describe damped oscillations of the output variable $\alpha(t)$ for a switching-off-step of the input current i. For $d > 1$, k_1 and k_2 are real and the output deflection $\alpha(t)$

will follow a steady decline towards zero. The case $d = 1$ will be found right in between the previous two cases. It need to be considered separately.

As there are two solutions for k, Euler's hypothetical expression for α needs to be modified. It should consist of two terms:

$$\dot{\alpha} = A_1 \exp(k_1 t) + A_2 \exp(k_2 t)$$

d < 1 (underdamped)

The starting conditions serve to determine A_1 and A_2. We may write:

$$k_{1,2} = \omega_0 (-d \pm j\sqrt{1-d^2})$$

With these values of k, we obtain the solution α as

$$\alpha = A_1 \exp(-d\omega_0 t) \cdot \exp(j\sqrt{1-d^2}\,\omega_0 t) + A_2 \exp(-d\omega_0 t)$$
$$\times \exp(-j\sqrt{1-d^2}\,\omega_0 t)$$
$$= \exp(-d\omega_0 t)(A_1 \exp(j\sqrt{1-d^2}\,\omega_0 t) + A_2 \exp(-j\sqrt{1-d^2}\,\omega_0 t)$$

Using Euler's formulae

$$\exp(\quad j\varphi) = \cos\varphi + j\sin\varphi$$
$$\exp(-j\varphi) = \cos\varphi - j\sin\varphi$$

we may transform the exponential functions into circular ones. They give a better understanding of the α-oscillations:

$$\alpha = \exp(-d\omega_0 t)\,[A_1\,(\cos\sqrt{1-d^2}\,\omega_0 t + j\sin\sqrt{1-d^2}\,\omega_0 t)$$
$$+ A_2\,(\cos\sqrt{1-d^2}\,\omega_0 t - j\sin\sqrt{1-d^2}\,\omega_0 t)]$$
$$= \exp(-d\omega_0 t)\,[(A_1 + A_2)\cos\sqrt{1-d^2}\,\omega_0 t + j(A_1 - A_2)\sin\sqrt{1-d^2}\,\omega_0 t)]$$

A_1 and A_2 are unknown. They will be determined using the starting conditions. For this purpose we need also

$$\frac{d\alpha}{dt} = \dot{\alpha} = -d\omega_0 \exp(-d\omega_0 t)\,[(A_1 + A_2)\cos\sqrt{1-d^2}\,\omega_0 t$$
$$+ j\,(A_1 - A_2)\sin\sqrt{1-d^2}\,\omega_0 t]$$
$$\exp(-d\omega_0 t)\,[-(A_1 + A_2)\sqrt{1-d^2}\,\omega_0\sin\sqrt{1-d^2}\,\omega_0 t +$$
$$+ j\,(A_1 - A_2)\sqrt{1-d^2}\,\omega_0\cos\sqrt{1-d^2}\,\omega_0 t]$$

The starting conditions are

$$\alpha(t=0) = \alpha_0 = A_1 + A_2$$
$$\dot{\alpha}(t=0) = 0 = -d\omega_0 \cdot \underbrace{(A_1 + A_2)}_{\alpha_0} + j\,(A_1 - A_2)\sqrt{1-d^2}\,\omega_0$$

The solution α contains $(A_1 + A_2)$ which is determined already as α_0. But we also need the expression of $j(A_1 - A_2)$ for α. It can be obtained from the last equation as:

$$j(A_1 - A_2) = \frac{d\omega_0}{\sqrt{1-d^2}\,\omega_0}\,\alpha_0$$

so the complete solution α may be written for this case of $d < 1$ as:

$$\frac{\alpha}{\alpha_0} = \exp(-d\omega_0 t)\left(\cos\underbrace{\sqrt{1-d^2}\,\omega_0 t}_{\omega} + \frac{d}{\sqrt{1-d^2}}\sin\underbrace{\sqrt{1-d^2}\,\omega_0 t}_{\omega}\right)$$

This function describes damped oscillations.

$d > 1$ (overdamped)

The solution of the differential equation was

$$\alpha = A_1\exp(k_1 t) + A_2\exp(k_2 t)$$

The constants k_1 and k_2 may be determined right away. They are the solutions of the characteristic equation. To obtain A_1 and A_2 the well known starting conditions may serve:

$$\alpha\,(t=0) = \alpha_0 \qquad\qquad \left(\frac{d\alpha}{dt}\right)_{t=0} = 0$$

From $\alpha(0)$ we obtain $\qquad \alpha_0 = A_1 + A_2$

From $\dfrac{d\alpha\,(0)}{dt}$, we obtain $\qquad 0 = A_1 k_1 + A_2 k_2.$

These are the two equations which allow to determine

$$A_1 = \alpha_0\frac{k_2}{k_2-k_1}; \quad A_2 = \alpha_0\frac{k_1}{k_1-k_2}$$

So we obtain the solution α for this case $d > 1$ as

$$\frac{\alpha}{\alpha_0} = \frac{k_2}{k_2-k_1}\exp(k_1 t) + \frac{k_1}{k_1-k_2}\exp(k_2 t)$$

This function describes a steady declining output for the switched off input. It should be used for drawing the graph. The constants k_1 and k_2 are determined from

$$k_{1,2} = \omega_0\left(-d \pm \sqrt{d^2-1}\right)$$

Of course k_1 and k_2 are always negative for $d > 1$, but if A_1 is positive, A_2 will be negative.

α/α_0 can be rewritten in the form derived under $d < 1$ by using hyperbolic functions

$$\cosh x = \frac{e^x + e^{-x}}{2}$$

$$\sinh x = \frac{e^x - e^{-x}}{2}$$

This will result as

$$\frac{\alpha}{\alpha_0} = \exp\left(-d\omega_0 t\right)\left[\cos h\underbrace{\left(\sqrt{d^2-1}\,\omega_0\right)}_{\omega} t + \frac{d}{\sqrt{d^2-1}}\sin h\underbrace{\left(\sqrt{d^2-1}\,\omega_0\right)}_{\omega} t\right]$$

In principle, the equations for α/α_0 are similar in both cases for $d \neq 1$.

d = 1 (aperiodically damped)

In this case the characteristic equation will be

$$k^2 + 2\omega_0 k + \omega_0^2 = 0$$

and we obtain that

$$k_1 = k_2 = k$$

$$k = -\omega_0.$$

The solution α will be found between the previous cases. The following statement allows to find it:

$$\alpha = A\left[(\exp\,(kt) - kt\,\exp\,(kt)\right]$$

$$= A\,(1 + \omega_0 t)\,\exp\,(-\omega_0 t).$$

As there is only one constant A to be determined we need just the first starting condition.

$$\alpha\,(t = 0) = \alpha_0 = A.$$

So we get the solution

$$\frac{\alpha}{\alpha_0} = (1 + \omega_0 t)\,\exp\,(-\omega_0 t).$$

For all three cases of d and also the special cases of critical damping $(d = 1/\sqrt{2})$ and of no damping $(d = 0)$ the graph of these functions is depicted in Fig. 2.2.1-1. In order to notice all solutions easily they are given again:

$$d < 1:\ \alpha = \alpha_0 \exp\left(-d\omega_0 t\right)\left[\cos\left(\sqrt{1-d^2}\,\omega_0\right)t\right.$$

$$\left. + \frac{d}{\sqrt{1-d^2}}\sin\left(\sqrt{1-d^2}\,\omega_0\right)t\right]$$

$$d > 1: \quad \alpha = \alpha_0 \exp\left(-d\omega_0 t\right) \left[\cosh\left(\sqrt{d^2-1}\,\omega_0\right) t \right.$$

$$\left. + \frac{d}{\sqrt{d^2-1}} \sinh\left(\sqrt{d^2-1}\,\omega_0\right) t \right]$$

$$d = 1: \quad \alpha = \alpha_0 \exp\left(-\omega_0 t\right)\left(1 + \omega_0 t\right)$$

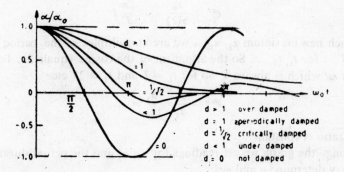

Fig. 2.2.1-1.   Step Response of a Second Order System (Switching Off-Behaviour)

Practical System Analysis

The practical system analysis deals with the determination of the system constants θ, D and C from the declining system oscillations. This is one goal. Another one is to get ω_0 and d which give $T_0 = 2\pi/\omega_0$ and $a = \omega_0 d$ as well. All these quantities describe the system behaviour perfectly. So let us switch off the steady current at the time zero. The system will respond with the decreasing deflection angle $a(t)$, Fig. 2.2.1-2, which allows the readings of T, $1/a$, α_0, α_1, α_2, etc.

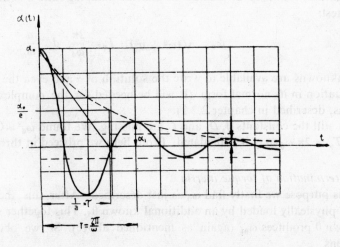

Fig. 2.2.1-2.   Damped Oscillations of a Second Order System.

The attenuation constant 'a' should be determined the following way because the reading of $1/a$ from the graph may suffer from considerable reading errors. We better use consecutive amplitudes which might be read quite accurately. From

$$\alpha(t) = \alpha_0 e^{-at} \cos \omega t$$

we obtain

$$e^{at} = \frac{\alpha_0}{\alpha(t)} \cos \frac{2\pi}{T} t$$

At each new maximum $\alpha_1, \alpha_2,...$ we are at multiples of the period duration $T, 2T,..$ for $t_1, t_2,....$ So the argument of the cosine equals $2\pi, 4\pi,...$ giving a cos ωt which is always 1. So for $t_1 = T$ and $t_2 = 2T$ etc.

$$e^{aT} = \frac{\alpha_0}{\alpha_1} = \frac{\alpha_1}{\alpha_2} = ...$$

This ratio is constant.

Using the graph we get readings for α_0, α_1 and the period duration T. So we may determine a and ω:

$$\ln(e^{aT}) = aT = \ln \frac{\alpha_0}{\alpha_1}$$

("Log. Decrement")

$$a = \frac{\ln \dfrac{\alpha_0}{\alpha_1}}{T} \text{ and } \omega = \frac{2\pi}{T}.$$

From these results of the attenuation constant 'a' and the circular frequency ω we may calculate the natural circular frequency ω_0, the period duration T_0 of natural (undamped) oscillations and the damping ratio d of the system under test:

$$\omega_0 = \sqrt{(\omega^2 + a^2)}; \quad T_0 = \frac{2\pi}{\omega_0}; \quad d = \frac{a}{\omega_0}$$

The unknowns are available to write the solution of α and even the differential equation in its normal form. (It will be needed for the complex system analysis, described in chapter 2.2.2).

But still the constants θ, D, and C are missing. We found $\omega_0^2 = C/\theta$ from which $T_0 = 2\pi/\omega_0$ may be calculated. From here we proceed in three steps.

1. *Determination of torque inertia* θ

For this purpose we firstly find ω_0 as just described. After this the system will be physically loaded by an additional known θ_1. This together with the unknown θ produces ω_{01} (again as mentioned above). So we obtain two equations

$$\omega_0^2 = \frac{C}{\theta} \qquad\qquad \omega_{01}^2 = \frac{C}{\theta + \theta_1}$$

The spring rate C (actually torque direction) is the same for the unloaded and the loaded case:

$$C = \theta\omega_0^2 \qquad C = (\theta + \theta_1)\,\omega_{01}^2$$

Comparing these equations we find that

$$\theta\omega_0^2 = (\theta + \theta_1)\omega_{01}^2$$

$$\theta(\omega_0^2 - \omega_{01}^2) = \theta_1\omega_{01}^2$$

$$\theta\left(\frac{\omega_0^2}{\omega_{01}^2} - 1\right) = \theta_1$$

$$\theta = \frac{\theta_1}{\left(\dfrac{\omega_0}{\omega_{01}}\right)^2 - 1}$$

2. *Determination of the spring rate C (torque direction)*
From $\omega_0^2 = C/\theta$ we obtain easily

$$C = \theta\omega_0^2$$

3. *Determination of the damping constant D*
We know from previous derivations that

$$D = 2d\sqrt{(\theta C)}$$

In order to determine the damping constant D the damping ratio 'd' needs to be determined first.

$$d = \frac{a}{\omega_0}$$

$$= \frac{\overbrace{\dfrac{\ln\dfrac{\alpha_0}{\alpha_1}}{T}}^{a}}{\underbrace{\sqrt{\underbrace{\left(\dfrac{2\pi}{T}\right)^2}_{\omega^2} + \underbrace{\left(\dfrac{\ln\dfrac{\alpha_0}{\alpha_1}}{T}\right)^2}_{a^2}}}_{\omega_0}}$$

Now the damping constant D can be determined.

2.2.2 Frequency Response of Second Order System
Complex Transfer Function and Bode Plot

A step input was only a special case for exciting the system under test. Quite the same a sinusoidal excitation can be applied and we may ask for the system response for changing input frequencies but of constant amplitude. As the second order system is linear the output produces the same frequency as applied to its input. But the output amplitude will be effected by the input frequency as well as the phaseshift between the input and output variables. The frequency response contains implicitly the step response, for a step may be described as the sum of harmonics, (see chapter 14: Fourier-analysis). Due to this fact all information so far obtained from the second order system will also be obtainable using the frequency response. For this purpose the transfer function $F(j\omega)$ will be derived. The damping ratio d may be obtained from it, and all other system constants as well, if needed.

We found the differential equation for the system under test. It serves to find the complex transfer function.

$$\theta \frac{d^2\alpha}{dt^2} + D \frac{d\alpha}{dt} + C\alpha = k_s i$$

$$\frac{\theta}{C} \frac{d^2\alpha}{dt^2} + \frac{D}{C} \frac{d\alpha}{dt} + \alpha = \frac{K_s}{C} i$$

The input variable was the current $i = X_i$, the output variable was the deflection angle $\alpha = X_0$. Similar to chapter 2.2.1 we use abbreviations:

$$\frac{\theta}{C} = \frac{1}{\omega_0^2}$$

$$D = 2\sqrt{(\theta C)}d$$

$$\frac{D}{C} = 2\sqrt{\left(\frac{\theta}{C}\right)}d = 2\frac{d}{\omega_0}$$

$$\frac{1}{\omega_0^2} \frac{d^2 X_0}{dt^2} + 2\frac{d}{\omega_0} \frac{dX_0}{dt} + X_0 = KX_I \quad \text{with} \quad K = \frac{K_s}{C}$$

Replacing (d/dt) by $j\omega$ (which is the Laplace-operator) and d^2/dt^2 by $(j\omega)^2$ quite formally, the frequency response $F(j\omega)$ will be directly obtained as ratio of the quantities X_0/X_i of output and input:

$$\left[\frac{(j\omega)^2}{\omega_0^2} + 2d\frac{j\omega}{\omega_0} + 1\right] X_0(j\omega) = KX_I \ (j\omega)$$

$$F(j\omega) = \frac{X_0 \ (j\omega)}{X_I \ (j\omega)} = \frac{K}{1 - \left(\frac{\omega}{\omega_0}\right)^2 + j2d\frac{\omega}{\omega_0}} = K\frac{\left[1 - \left(\frac{\omega}{\omega_0}\right)^2\right] - 2jd\frac{\omega}{\omega_0}}{\left[1 - \left(\frac{\omega}{\omega_0}\right)^2\right]^2 + \left[2d\frac{\omega}{\omega_0}\right]^2}$$

All variables depend on frequency now.

For a certain frequency of X_i the performance of X_0 may be obtained from $X_0 = F \cdot X_i$.

Usuallay the complex function of $F(j\omega)$ is depicted as a graph of its magnitude and its phase, each over the frequency (Fig. 2.2.2-1) giving the Bode-plot of the second order system.

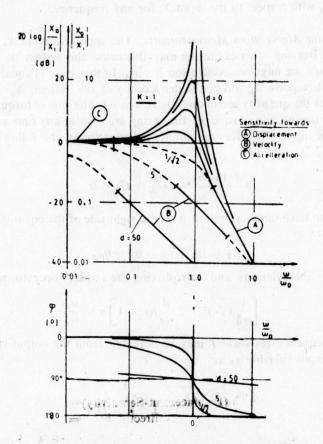

Fig. 2.2.2-1. Bode Plot of the Second Order System.

$$\text{Magnitude} \quad \left|\frac{x_0}{x_i}\right| = \frac{K}{\sqrt{\left[- \left(\frac{\omega}{\omega_0}\right)^2\right]^2 + \left[2d\frac{\omega}{\omega_0}\right]^2}} = f\left(\frac{\omega}{\omega_0}\right)$$

$$\text{Phase} \quad \tan\varphi = \frac{I_m\left[\dfrac{x_0}{x_i}\right]}{Re\left[\dfrac{x_0}{x_i}\right]}$$

$$\varphi = \text{arc tan} \; \frac{-2\,d\dfrac{\omega}{\omega_0}}{1 - \left(\dfrac{\omega}{\omega_0}\right)^2} = f\left(\frac{\omega}{\omega_0}\right)$$

The magnitude graph usually serves to determine the amplitudes of the output X_0 with respect to the input X_i for any frequency.

Ex. *Angular Acceleration Measurements.* The input variable X_i was the current I. But any other excitation may also cause the system to respond; for instance an angular acceleration $\ddot{\beta} = a_A$. In this case X_i equals a_A and the output response X_0 will be the deflection α of the system, $X_0 = \alpha$. This means that the quantity α will directly be sensed. The sum of torque inertia, damping torque and torque of the spring are zero at any time as there is no current input. This gives the differential equation in the following way, (Fig. 2.2.2-2).

$$\theta \frac{d^2\,(\beta-\alpha)}{at^2} - D\frac{d\alpha}{dt} - C\alpha = 0.$$

In order to have the acceleration $\ddot{\beta}$ on the right side of the equation, it may be rewritten as:

$$\theta\ddot{\alpha} + D\dot{\alpha} + c\alpha = \theta\ddot{\beta} = \theta a_A$$

Using the abbreviations and introducing the Laplace-operator we obtain

$$\left[\frac{1}{\omega_0^2}\,(j\omega)^2 + 2\frac{d}{\omega_0}\,j\omega + 1\right]\alpha = \frac{a_A}{\omega_0^2}$$

So the frequency response F may be obtained from the output variable α over the input variable a_A as:

$$F = \frac{\alpha}{a_A} = \frac{1}{\omega_0^2}\frac{1}{1 - \left(\dfrac{\omega}{\omega_0}\right)^2 + j2\,d\,\dfrac{\omega}{\omega_0}}$$

From that the frequency dependence of the magnitude $\left|\dfrac{\alpha}{a_A/\omega_0^2}\right|$ will be the same as shown before and as depicted in Fig. 2.2.2-1. The output α suffers from a phaseshift φ with respect to the input $\ddot{\beta} = a_A$. Three cases can be highlighted for practical use.

Case A: D and C are Small (Displacement-Sensitivity)
From the differential equation it follows directly that $\ddot{\alpha} = \ddot{\beta}$, or $\alpha = \beta$. $\ddot{\beta}$ was the angular acceleration to be measured, which means that β is the angular displacement. As the device responds directly in the form of α, the system is "displacement-sensitive" for the case A. But as the angular acceleration $\ddot{\beta}$ should be measured, two differentiations of the output quantity α need to be effected.

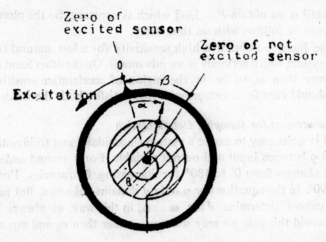

Fig. 2.2.2-2 Damped Mass-Spring-System with External Excitation β

As C is small the natural frequency will be low, see first abbreviation. So the frequency region for which the above mentioned statements are valid will be found for $\omega \gg \omega_0$. We can say that the system is low tuned. The slope of the magnitude will be -40 dB per decade frequency change as follows from $F(j\omega)$, because we obtain $\alpha = -a_A/\omega^2$ for this ω-range.

Using the formula for φ we obtain a phase shift between output deflection α and input displacement β which is $\varphi = 0$. (But $a_A = \beta$ of the actual system is opposite to α. So in reality we have 180° phase shift which was already taken into account in Fig. 2.2.2.-1). Anyway φ equal to zero means that the angular displacement measurement takes place without time delay.

Case B: θ and C are small (Velocity-Sensitivity)
Using the differential equation we obtain $D\dot\alpha = \theta\ddot\beta$, or $\alpha = \theta\dot\beta/D$. As $\dot\beta$ is the angular velocity we say the system is "velocity-sensitive". But as the angular acceleration $\ddot\beta$ should be measured, a differentiation of the output α needs to be effected.

As the damping effects the system behaviour, the slope of the magnitude will be -20 dB per decade frequency change, which follows from $F(j\omega)$, because $\alpha = \dfrac{1}{2d\omega_0}\dfrac{a_A}{j\omega}$. For $\omega = \omega_0$ the phase gets 90° for any damping.

Case C: θ and D Small (Acceleration-Sensitivity)
The differential equation gives $C\alpha = \theta\ddot\beta$, that means

$$\alpha = \frac{\theta}{C}\ddot\beta = \ddot\beta/\omega_0^2.$$

The sensed output variable α is directly proportional to the angular acceleration $\ddot\beta$ to be measured. The system is "acceleration-sensitive". This is the case for low exciting frequencies ω, especially for $\omega \ll \omega_0$. We have a high tuned system.

For small ω we obtain $F = 1/\omega_0^2$ which is constant. So the phase will be equal to zero. α follows without time delay.

On one hand we obtain a high sensitivity for a low natural frequency ω_0 of the system which certainly is an advantage. On the other hand ω should be far lower than ω_0 to be in the region of acceleration sensitivity. The engineer should care for a compromise that satisfies both the needs.

Phase-measurement for Damping Determination

Usually it is quite easy to excite a system sinusoidally and to investigate the phaseshift φ between input and output variable. For a second order system the phase changes from $0°$ to $180°$ for increasing frequencies. For $\omega = \omega_0$ it will be $90°$. In the equation for φ only d remains unknown. But in spite of that we cannot determine d for $\omega = \omega_0$ in this way, as always becomes zero. To avoid this case we may use any ω other than ω_0 and measure φ to obtain

$$d = \frac{-\tan \varphi}{2\frac{\omega}{\omega_0}}\left[1 - \left(\frac{\omega}{\omega_0}\right)^2\right]$$

Again: D may be determined now from

$$D = 2d\sqrt{\theta C}$$

The described procedure may easily be followed in practice. Only φ and ω readings are necessary. In order to show that, quite a complex theoretical handling was necessary.

But d could also be obtained from the magnitude graph, for $\tan \varphi$ and the slope of the magnitude have a fixed relation to each other. To show this relation the first derivative of the magnitude with respect to ω/ω_0 needs to be calculated. We will not deal with this case.

2.2.3.   Damping of a Galvanometer

For sensitive measurements galvanometers are appropriate. In order to provide a very light system weight of the moving parts the damping is provided externally with the help of R_e. (Fig. 2.2.3-1.) Using taut strips as suspension, ensures a very low spring rate. The damping behaviour may be studied for a switched-off measurand I. At $t = 0$, the coil starts to go back to zero position. Due to its velocity v it generates a voltage E which in turn drives the current I_D through R_e and the internal coil resistance R_i. This current I_D produces a reacting torque T_D which damps the movement. (See also Fig. 1-2).

$$E = Nl\,(B \times v) \quad E = NlBv \quad l = 2h$$

$$v = \frac{dc}{2}\frac{d\alpha}{dt} = N2hB\frac{d_c}{2}\frac{d\alpha}{dt}$$

$$[d_c = \text{coil diameter}]$$

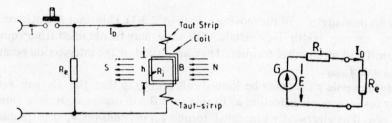

Fig. 2.2.3-1. External Galvanometer Damping.

$$I_D = \frac{E}{R_i + R_e} = \frac{NhBd_c}{R_i + R_e}\frac{d\alpha}{dt}$$

$$T_D = 2F\frac{d_c}{2}$$

$$\mathbf{F} = Nh\,(I_D \times B) \qquad F = NhBI_D$$

$$= \underbrace{\frac{(NhBd_c)^2}{R_i + R_e}}_{D}\frac{d\alpha}{dt}$$

Looking back to chapter 2.2.1 we find one torque which was proportional to the angular velocity $d\alpha/dt$. The proportionality factor was the damping factor D which may be determined with the help of the external resistor R_e. As D is known the relative damping coefficient d may now be calculated from

$$d = \frac{1}{2}\frac{D}{\sqrt{C\theta}}$$

Usually $d = 1/\sqrt{2}$ is realized which can be effected by R_e.

2.3 THE SCALE OF DIFFERENT MOVING COIL SYSTEMS LINEAR AND NON LINEAR SCALES

All moving coil systems are normally based on the action of a force generated by the magnetic field of the moving coil due to the measured current and the magnetic field of a permanent magnet. The magnetic path consists of the magnet, of soft iron paths to guide the flux, and of pole pieces to shape the field inside the air gaps.

In chapter 1.1.3 the scale equation was developed. As we deal with scales now, this equation should be solved for the scale deflection

$$\alpha = \frac{Nh\,d_cB}{C}\cdot I$$

For equal increases of the current, equal increases of the deflection will take place. Therefore the scale is linear if the flux density B is the same for all deflection angles α of the moving coil (Fig. 2.3-1). Constant B is achieved as the air gap width is constant. The scale may be obtained from equal electrical and mechanical torques. They are found at the intersection points of T_{el} and T_{mech}.

But the pole pieces may be shaped differently, so that the air gap gets wider for increasing deflection α. In this case B will decrease. It is a function of α. The electrically generated torque T_{el} depends on α. The pointer deflection α will be found at the equilibrium position for which T_{el} is compensated by the mechanical torque T_{mech} of the spring. The scale is compressed in its upper range. The instrument provides high sensitivity in its lower range and low sensitivity in its upper range.

This type of mechanical design of Fig. 2.3-1 is called outer magnet system. It suffers from considerable stray flux. For very sensitive instruments the magnet needs a large volume to generate high flux density. The core magnet system of Fig. 2.3-2 avoids this disadvantage. It may be small, still providing high sensitivity. It is used to achieve compressed scale divisions either in the upper or in the lower scale range.

This configuration is signified by a magnetic field that follows a cosine function and consequently the electric torque T_{el} follows as well:

$$B(\alpha) = \hat{B} \cos \underbrace{(\alpha - \beta)}_{\delta}.$$

The angle β is the angular distance of the zero position from the field strength maximum (for the coil windings). β may be chosen freely thus providing the possibility to compress the scale either in its upper or lower range. The scale equation may be achieved in an implicit form employing the equilibrium condition of the acting torques:

$$T_{el} = T_{mech}$$

$$IN\, hd_c\hat{B} \cos (\alpha - \beta) = C\alpha$$

$$I = \frac{C}{Nh\, d_c\hat{B}} \cdot \frac{\alpha}{\cos (\alpha - \beta)}$$

The intersection points of T_{el} and T_{mech} provide the scale, see Fig. 2.3-2.

Chosing β the spring rate C will be determined from the last equation. For this purpose we fix the maximum current I_{max} which should produce full scale deflection $\alpha_{fsd} = 90°$.

$$C = I_{max}\, Nh\, d_c\hat{B}\, \frac{\cos (\alpha_{fsd} - \beta)}{\alpha_{fsd}}$$

For an angle β_1, Fig. 2.3-2 gives the mechanical torque T_{mech_1}. It is zero for $\delta = -\beta_1 (\alpha_1 = 0)$. To fix the straight line of T_{mech_1} a second point is

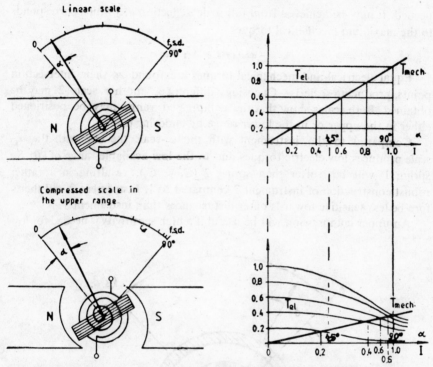

Fig. 2.3–1.   Linear and Non Linear Scales of Moving Coil
Instruments with Outer Magnet.

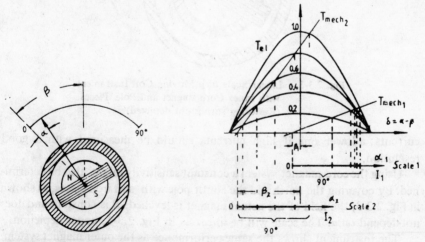

Scale 1:   Compressed upper range
Scale 2:   Compressed lower range

Fig. 2.3-2.   Scales of Moving Coil Instruments with Inner
Core Magnet. (Restoring spring not depicted)

needed. It may be achieved from full scale deflection α_{fsd} which corresponds to the maximum mechanical torque

$$T_{mech,\,fsd} = C_1\,\alpha_{fsd}$$

Equal electrical and mechanical torques are found as their intersection points, producing scale 1.—Choosing a different β_2 another scale 2 may be obtained. Both scales show that the compressed range may be positioned either in the upper or in the lower range by choosing β.

Compared to the instrument with the α_2-scale the one with the α_1-scale produces low electric torques due to the low available range of B. So spring 1 will be softer than spring 2 $(C_1 < C_2)$, resulting in a rather robust construction of instrument 2 compared to 1. Instrument 2 will therefore be less sensitive towards outer disturbances than instrument 1.

An upper compression will be useful if a high sensitivity is needed for low

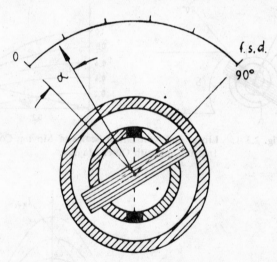

Fig. 2.3-3.   Linear Scale of a Moving Coil Instrument
with Inner Core Magnet and Pole Pieces.
(Restoring spring not depicted)

currents; a lower one if high currents should be measured with a good resolution.

Using the core magnet system a constant sensitivity might also be determined. By covering the north and the south pole with soft iron pieces, as shown in Fig. 2.3-3, the field of the actual magnet is levelled. B is constant and does not depend on α. The scale will be linear as in Fig. 2.3-1, upper depiction.

The instrument shows the same performance as the outer magnet system. As the stray flux is very small the volume of the magnet may be small. Core magnet instruments are very small. The outer ring serves mainly to guide the flux, and also to shield the system against external disturbing fields. This is another point facilitating high sensitivity of inner core assemblies.

Moving coil types are the most sensitive deflection instruments. Galvanometers may sense 10 μA and still produce full scale deflection. The range up to 50 mA may easily be covered. Due to the torque inertia of the movable parts they are preferably used for *DC*. By employing additional rectifiers *AC* may also be sensed. Moving coil instruments cover the accuracy classes ranging from 0.1 to 5.

2.4 MULTI RANGE METER

Ampere and volt meters are expected to cover wide ranges of currents and voltages. In order to obtain a good deflection of the indicator for each range, multi range meters have been developed. At the same time they are usually equipped as ohm-meters.

2.4.1. Ammeter

Moving coil systems are usually manufactured for a full scale deflection for 10 μA to 50 mA. Currents higher than these cannot be measured directly due to technical construction limitations. There is firstly the size of the coil which should not exceed certain measures. Secondly the self heating of the coil due to its power dissipation should be low to avoid temperature effects. Thirdly the leads which are usually the spiral springs and the bearings pose mechanical limitations. But higher currents may be measured employing shunt resistors. They form a bypass and only a part of the actual current to be measured will be allowed to pass through the moving coil.

The ammeter is connected in series with the load. The voltage drop across the instrument should be low (2. .200mV). We may substitute the real ammeter by two ideal components. They are the actual instrument, now considered to cause no losses and the internal resistance R_M of the meter, (Fig. 2.4.1-1). Instead of R_M the voltage drop U_M across R_M may be labelled for full scale deflection current I_{fsd}.

Shunting

If the measured current is greater than I_{fsd}, the instrument needs to be shunted, using R_S.

Ex. Let us use a 30 mA instrument having an internal resistance of $R_M = 22$ ohms. A shunt resistor of $R_S = 11$ ohms is applied. (Fig. 2.4.1-1). If a current of 30 mA is applied only 10 mA will be indicated. 20 mA will pass through R_S. This way we have increased the measurement range by a factor of 3. Now 90 mA would be measurable for full scale deflection. Note: By means of shunt resistors an ammeter may cover wide ranges of measurable currents.

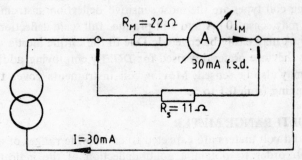

30 mA reading directly
10 mA reading with R_S.

Fig. 2.4.1-1. Shunting an Ammeter.

Technical Realization of a Multi-Range Ammeter
The shunting is usually effected by a circuit shown in Fig. 2.4.1-2.

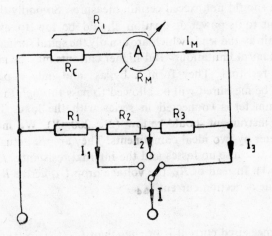

Fig. 2.4.1-2. Current Divider Circuit of a Multi Range
Ammeter.

The internal resistance R_i consists usually of the winding resistance R_M and the component R_C for temperature compensation (see chapter 2.5). The range selector switch, connected in this way, avoids reading changes for changing contact resistance.

Calculation of the Current Divider Circuit
The voltage relations may be used to determine the scaling resistors R_1, R_2 and R_3:

$$R_i I_M = U_M = (I_3 - I_M) R_1 + (I_3 - I_M) R_2 + (I_3 - I_M) R_3 \quad \text{for range 3}$$

$$U_M = (I_2 - I_M)\, R_1 + (I_2 - I_M)\, R_2 - \qquad I_M\ R_3 \quad \text{for range 2}$$

$$U_M = (I_1 - I_M)\, R_1 - \qquad I_M\ R_2 - \qquad I_M\ R_3 \quad \text{for range 1}$$

From that the matrix M and the determinants ΔA, ΔB_1, ΔB_2, and ΔB_3 may be obtained.

$$M = \begin{pmatrix} U_M \\ U_M \\ U_M \end{pmatrix} = \underbrace{\begin{pmatrix} I_3 - I_M & I_3 - I_M & i_3 - I_M \\ I_2 - I_M & I_2 - I_M & -I_M \\ I_1 - I_M & -I_M & -I_M \end{pmatrix}}_{\Delta A} \begin{pmatrix} R_1 \\ R_2 \\ R_3 \end{pmatrix}$$

$$\Delta B_1 = \begin{vmatrix} U_M & I_3 - I_M & I_3 - I_M \\ U_M & I_2 - I_M & -I_M \\ U_M & -I_M & -I_M \end{vmatrix}$$

$$\Delta B_2 = \begin{vmatrix} I_3 - I_M & U_M & I_3 - I_M \\ I_2 - I_M & U_M & -I_M \\ I_1 - I_M & U_M & -I_M \end{vmatrix}$$

$$\Delta B_3 = \begin{vmatrix} I_3 - I_M & I_3 - I_M & U_M \\ I_2 - I_M & I_2 - I_M & U_M \\ I_1 - I_M & -I_M & U_M \end{vmatrix}$$

The unknown scaling resistors are

$$R_1 = \frac{\Delta B_1}{\Delta A}; \quad R_2 = \frac{\Delta B_2}{\Delta A}; \quad R_3 = \frac{\Delta B_3}{\Delta A}.$$

The dimension of $[\Delta A] = 1A^3$, the one of $[\Delta B_n] = 1VA^2$. So the dimension of $[R_n] = 1\Omega$.

The calculation procedure of this approach is recommended for practical use and simplicity reasons. Another approach may be based on the current relations of the circuit. But the calculations turn out to be quite complicated

Ex. A moving coil ammeter should be used to assemble a multi meter for the current ranges of

$$I_3 = 2\,\text{mA}, \ I_2 = 20\,\text{mA}, \ I_1 = 200\,\text{mA}.$$

The instrument provides full scale deflection for $I_M = 1\text{mA}$. Its internal resistance is $R_1 = 10\Omega$. Determine the scaling resistors R_1, R_2 and R_3.

$$U_M = I_M \cdot R_i = 1\,\text{mA} \cdot 10\Omega = 10\,\text{mV}.$$

$$M = \underbrace{\begin{pmatrix} 1\,\text{mA} & 1\,\text{mA} & 1\,\text{mA} \\ 19\,\text{mA} & 19\,\text{mA} & -1\,\text{mA} \\ 199\,\text{mA} & -1\,\text{mA} & -1\,\text{mA} \end{pmatrix}}_{\Delta A} \begin{pmatrix} R_1 \\ R_2 \\ R_3 \end{pmatrix} = \begin{pmatrix} 10\,\text{mV} \\ 10\,\text{mV} \\ 10\,\text{mV} \end{pmatrix}$$

$$\Delta A = \begin{vmatrix} 1\,\text{mA} & 1\,\text{mA} & 1\,\text{mA} \\ 19\,\text{mA} & 19\,\text{mA} & -1\,\text{m}\overset{\text{A}}{} \\ 199\,\text{mA} & -1\,\text{mA} & -1\,\text{mA} \end{vmatrix} = -4000 \cdot 10^{-9}\,\text{A}^3$$

$$\Delta B_1 = \begin{vmatrix} 10\,\text{mV} & 1\,\text{mA} & 1\,\text{mA} \\ 10\,\text{mV} & 19\,\text{mA} & -1\,\text{mA} \\ 10\,\text{mV} & -1\,\text{mA} & -1\,\text{mA} \end{vmatrix} = -400 \cdot 10^{-9}\,\text{VA}^2$$

$$\Delta B_2 = \begin{vmatrix} 1\,\text{mA} & 10\,\text{mV} & 1\,\text{mA} \\ 19\,\text{mA} & 10\,\text{mV} & -1\,\text{mA} \\ 199\,\text{mA} & 10\,\text{mV} & -1\,\text{mA} \end{vmatrix} = -3600 \cdot 10^{-9}\,\text{VA}^2$$

$$\Delta B_3 = \begin{vmatrix} 1\,\text{mA} & 1\,\text{mA} & 10\,\text{mV} \\ 19\,\text{mA} & 19\,\text{mA} & 10\,\text{mV} \\ 199\,\text{mA} & -1\,\text{mA} & 10\,\text{mV} \end{vmatrix} = -36000 \cdot 10^{-9}\,\text{VA}^2$$

The unknowns result as:

$$R_1 = \frac{\Delta B_1}{\Delta A} = \frac{-400 \cdot 10^{-9}\,\text{VA}^2}{-4000 \cdot 10^{-9}\,\text{A}^3} = 0.1\,\Omega$$

$$R_2 = \frac{\Delta B_2}{\Delta A} = \frac{-3600 \cdot 10^{-9}\,\text{VA}^2}{-4000 \cdot 10^{-9}\,\text{A}^3} = 0.9\,\Omega$$

$$R_3 = \frac{\Delta B_3}{\Delta A} = \frac{-36000 \cdot 10^{-9}\,\text{VA}^2}{-4000 \cdot 10^{-9}\,\text{A}^3} = 9.0\,\Omega$$

Proof. In the range of $I_3 = 2\,\text{mA}$ for full scale deflection the sum of

$$R_1 + R_2 + R_3 = R_i$$

because $I_M = 1\,\text{mA}$. Indeed $(0.1 + 0.9 + 9.0)\Omega = 10\Omega$ as for R_i.

2.4.2 Voltmeter

A moving coil instrument is actually sensitive to the current, shown in chapter 2.1. But due to Ohm's law we may measure a voltage, too. It only needs to be transferred into a current which is usually done with the help of a series resistor. If this resistor can be switched over to other values different ranges of the meter are obtainable.

Series Resistors

A moving coil ammeter is used. The series resistor R_V changes it to a voltmeter, (Fig. 2.4.2–1). The deflection α will be proportional to the current I_M, which can be expressed by the voltage U to be measured.

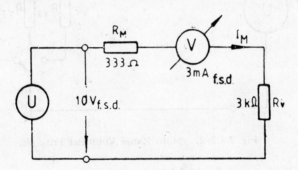

Fig. 2.4.2–1. Voltmeter.

$$\alpha = C_I \cdot I_M \qquad I_M = \frac{U}{R_V + R_M}$$

$$\alpha = \frac{C_I}{R_V + R_M} \cdot U$$

This is the scale equation of the voltmeter. The instrument is activated by the current I_M, but calibrated in terms of the measured voltage U. Note: The moving coil voltmeter is an ammeter calibrated in volts.

Ex. Let us use a 3mA-instrument with an internal resistance $R_M = 333\,\Omega$ and calculate the series resistor R_V to obtain full scale deflection for 10 V.

$$U = (R_V + R_M) \cdot I_M$$

$$R_V = \frac{U}{I_M} - R_M = \frac{10\text{V}}{3\text{mA}} - 333\Omega = 3000\Omega.$$

The additional resistor R_V increases the total internal resistance $R_i = R_V + R_M$. This way the loading effect is reduced considerably. As R_i depends on the voltage range the internal resistance of a voltmeter is usually given as ohms per volt. In the last example it would be

$$\frac{R_i}{U_{fsd}} = \frac{R_V + R_M}{U_{fsd}} = \frac{(3000 + 333)\Omega}{10V} = 333.3 \frac{\Omega}{V}$$

To cover wide voltage ranges different series resistors are employed. They are actuated by a selector switch. As the instrument current I_M is usually small, the contact resistance and changes of it need not to be considered, (Fig. 2.4.2-2).

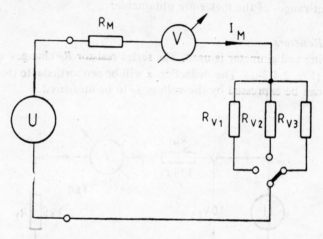

Fig. 2.4.2–2.   Multi Range Voltmeter Principle.

Technical Realization of a Multi-Range Voltmeter
An actual instrument is shown in Fig. 2.4.2–3. Instead of having three

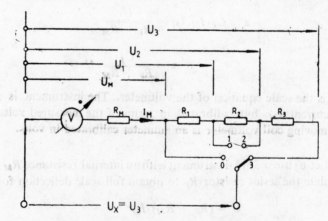

Fig. 2.4.2–3.   Multi Range Voltmeter.

different branches of series resistors for each range the resistor R_1 is also used in the range 2, and $(R_1 + R_2)$ are used in the range 3. This gives the advantage that R_2 and R_3 may be of smaller dissipation loss capability than the series resistors of the previous circuit.

Calculation Equations
For determination of the series resistances R_1, R_2, and R_3, the following voltage relations are employed:

$$\frac{R_M}{R_1} = \frac{U_M}{U_1 - U_M} \curvearrowright R_1 = \frac{U_1 - U_M}{U_M} R_M$$

$$\frac{R_M}{R_2} = \frac{U_M}{U_2 - U_1} \curvearrowright R_2 = \frac{U_2 - U_1}{U_M} R_M$$

$$\frac{R_M}{R_3} = \frac{U_M}{U_3 - U_2} \curvearrowright R_3 = \frac{U_3 - U_2}{U_M} R_M$$

Generally we find the nth resistor for n ranges

$$R_n = \frac{U_n - U_{n-1}}{U_M} R_M$$

A successive procedure is followed to determine R_n. U_M may be calculated first as $U_M = I_M \cdot R_M$. Together with the full scale voltage U_1, the value of R_1, for range 1 is obtained. The full scale voltage U_2 allows to fix R_2, for range 2 etc. for all R_n.

2.4.3 Ohm Meter

Ohm meters, as they are usually assembled in multi-range instruments, employ the circuit of Fig. 2.4.3-1. The unknown resistor Rx is inserted into the current circuit of a moving coil instrument. The current to be indicated is obtained from a battery which is actuated by a push button during the measurement.

Ex. 1.5 V are available. But only 1 V is used. After shorting Rx this voltage can be set to 1 V by changing the slider position of the potentiometer as to reach full scale deflection. As the shorting blank represents an $Rx = 0$ full scale deflection means zero for the resistance instrument. An open circuit at the input terminals means $Rx = \infty$. No current will flow. So there is no deflection.

For the instrument as shown in Fig. 2.4.3-1 the calculations have been executed. The deflection α is caused by the current I_M which is determined by the measurand Rx.

$$\alpha = C_I \underbrace{\frac{1V}{500\Omega + Rx}}_{I_M}$$

This is the scale equation of the Ohm-meter, which is a hyperbola, see Fig. 2.4.3-1 (b). The Ohm-scale is extremely non linear. It is depicted in Fig. 2.4.3-1 (c).

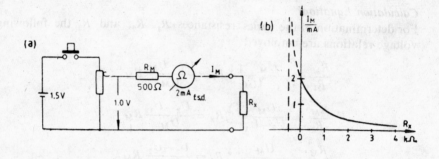

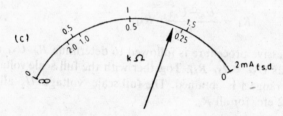

Fig. 2.4.3-1   Ohm-meter: (a) Circuit.   (b) Current-Hyperbola.   (c) Scale.

A multi-range Ohm-meter may be obtained using the circuit of Fig. 2.4.3-2.

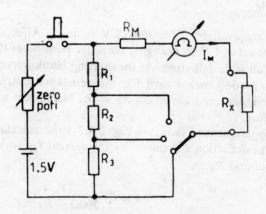

Fig. 2.4.3-2.   Multi Range Ohm-meter.

2.5 TEMPERATURE COMPENSATION OF THE COIL RESISTANCE

Instrument coils are made from copper wire which suffers from a temperature coefficient of its resistance as $\gamma_{cu} = 4 \cdot 10^{-3}/K$. Within applicable temperatures T, the dependence of the coil resistance R_M from T will be linear:

$$R_M = f(T) = R_{M0}(1 + \gamma_{Cu}\,\Delta T) = R_{M0} + \Delta R_M$$

$\Delta T = [T - T_o]$ gives the actual temperature deviation from normal temperature T_0.

For temperature compensation purposes a constant resistor R_C made from constantan is inserted, to obtain a voltmeter reading which is considerably less dependent on the environment temperature T, than the instrument without R_C, (Fig. 2.5-1.)

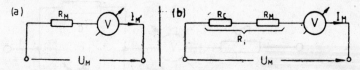

Fig. 2.5-1.   Voltmeter: (a) uncompensated (b) compensated.

Actually the reading α of the instrument is caused by the current I_M: $\alpha = C_1 I_M$. In this case not I_M but the voltage U_M is measured, and so U_M needs to be introduced.

$$\alpha = C_I \frac{U_M}{R}$$

$$\alpha_a = C_I \frac{U_M}{R_M} \qquad\qquad \alpha_b = C_I \frac{U_M}{R_C + R_M}$$

But R_M suffers from changes ΔR_M due to temperature changes which cause reading changes $\Delta\alpha$

$$\Delta\alpha = \frac{\partial\alpha}{\partial R_M}\,\Delta R_M$$

$$\Delta\alpha_a = -\frac{C_I U_M}{R_M{}^2}\,\Delta R_M \qquad\qquad \Delta\alpha_b = -\frac{C_I U_M}{(R_C + R_M)^2}\,\Delta R_M$$

$$\frac{\Delta\alpha_a}{\alpha_a} = -\frac{C_I U_M}{R_M{}^2}\,\Delta R_M\,\frac{R_M}{C_I U_M} \qquad\qquad \frac{\Delta\alpha_b}{\alpha_b} = -\frac{1}{R_C + R_M}\,\Delta R_M$$

$$= -\frac{\Delta R_M}{R_M} \qquad\qquad\qquad\qquad = -\frac{1}{\dfrac{R_C}{R_M} + 1}\,\frac{\Delta R_M}{R_M}$$

$$\frac{\Delta\alpha_b}{\alpha_b} = \frac{\Delta\alpha_a}{\alpha_a} \cdot \frac{1}{1 + \dfrac{R_C}{R_M}}$$

The comparison of the relative reading changes $\Delta\alpha/\alpha$ for the uncompensated instrument (a) and the compensated one (b) shows that the temperature dependence for (b) is less than that for (a) by a factor of $1/(1 + R_c/R_M)$. To obtain a considerable compensation effect R_c should be much greater than R_M. But unfortunately this measure decreases the sensitivity of the instrument at the same time. Usually a compromise is realized as $R_c = 10 \ R_M$.

The original sensitivity can nearly be maintained by employing an additional non linear NTC-element R_T called thermistor. The circuit is shown in Fig. 2.5-2. Further details on thermistors will be given in chapter 16.7.3.

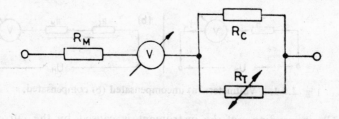

Fig. 2.5-2. Temperature Compensation of a Voltmeter without Essential Loss of Sensitivity.

The temperature effect can easily be demonstrated by heating up the coil of a voltmeter. As the coil resistance R_M determines the current I_M to which the instrument responds, a change ΔR_M certainly causes an indication change $\Delta\alpha$.

For an ammeter this is not the case. The measured current I is determined by the load, because the load resistance R_L is usually much higher than the internal resistance R_M of the ammeter. So, if a temperature change effects a change ΔR_M, this does not cause a considerable current change. I remains nearly constant and so the reading α is constant as well.

2.6 CROSS COIL INSTRUMENTS (RATIO-METERS)

2.6.1 Design and Features

Cross coil instruments can measure the ratio of two currents. For this purpose two moving coils are rigidly fixed together and mounted on the same shaft. They are fed by the currents I_1 and I_2. Small soft gold strips serve as leads. They do not provide any restoring torque. In fact no control torque, at all, should effect the movement of the coils.

The magnetic field is usually provided by a permanent magnet. It provides an induction B which depends on the cosine of the deflection angle $\alpha : B(\alpha) = \hat{B} \cos \alpha$.

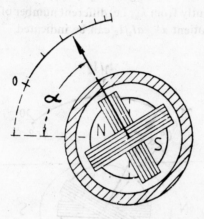

Fig. 2.6.1-1.   Principle of a Cross Coil Instrument.

The torques of the two coils are opposite to each other, so one serves to control the other. If the cross coil angle equals 90° the coils produce the torque T_1 and T_2.

$$T_1 = \underbrace{N_1 h d_c \hat{B}}_{k_1} \cdot I_1 \cos \alpha$$

$$T_2 = \underbrace{N_2 h d_c \hat{B}}_{k_2} \cdot I_2 \sin \alpha$$

(The same smybols for the quantities are used as for the moving coil instrument already described in chapter 1.1.3). The indication angle α is obtained from the equilibrium condition $T_1 = T_2$:

$$k_1 I_1 \cos \alpha = k_2 I_2 \sin \alpha$$

$$\underbrace{\frac{k_1}{k_2}}_{k} \frac{I_1}{I_2} = \frac{\sin \alpha}{\cos \alpha} = \tan \alpha$$

$$\alpha = \text{arc tan} \left(k \frac{I_1}{I_2} \right).$$

This is the scale equation of the ratio meter. It shows a non-linearity of the scale with respect to the measured current ratio I_1/I_2. But linearity can be obtained by providing a smaller cross coil angle and by composing the air gap wider for the pole region than for regions away from the poles. (Fig. 2.6.1-2). Thus the induction $B(\alpha)$ is shaped in such way as to obtain a linear scale. Usually $k_1 = k_2$, so the scale equation reads as:

$$\alpha = \frac{I_1}{I_2}$$

In case one current is to be weighted by a factor '*a*' this can be effected by designing k_1 differently from k_2, i.e. different number of windings: $N_1 = aN_2$. In this case the quotient $\alpha = aI_1/I_2$ can be indicated.

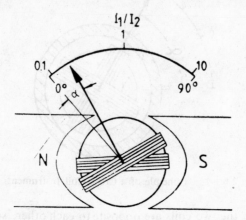

Fig. 2.6.1-2.   Real Cross Coil Instrument with Outer Magnet System.

A smaller cross coil angle provides the feature that an indication range of 90° can be covered.

2.6.2   Applications of Cross Coil Instruments

Ratio meters provide the advantage of not depending on the supply voltage or on other disturbing influences, especially temperature effects.

Ohm Meter
Fig. 2.6.2-1 shows the circuit. The battery voltage U_B is common for both the branches.

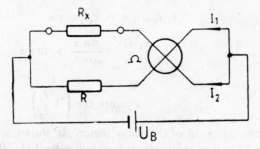

Fig. 2.6.2-1.   Ohm Meter Employing a Cross Coil Instrument.

$$U_B = R_X \cdot I_2 \quad \text{and} \quad U_B = R \cdot I_1$$

The left sides of these equations are equal, so the right sides are equal, too. We can solve them for the quotient I_1/I_2 which was equal to the indication α:

$$\alpha = \frac{I_1}{I_2} = \frac{R_X}{R}$$

The scale equation $R_X(\alpha)$ shows that R_X can directly be obtained from the reading α:

$$R_X = R \cdot \alpha$$

The scale of such ohm meter will be linear. Therefore this resistance measurement method is superior to the one described in chapter 2.4.3. Also changes of U_B will not effect the result. An **AC** supply might even be used.

Position Measurement Gauge (3 wire circuit)
The advantages of the cross coil instrument may be used to measure the angular position of a potentiometer over a long distance, see Fig. 2.6.2-2.

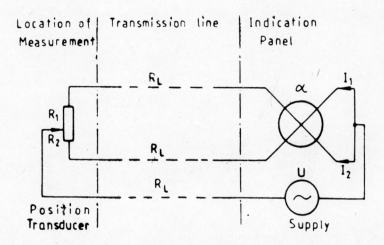

Fig. 2.6.2-2.   3-Wire Circuit for Position Measurement on Long Distances.

$$\alpha = \frac{I_1}{I_2}$$

$$= \frac{U}{R_2 + 2R_L} \frac{R_1 + 2R_L}{U}$$

Especially for supervision of the specified position, for which $R_1 = R_2$, this circuit will be applicable. This means: the middle position of the slider should preferably be monitored. In this case the currents I_1 and I_2 are equal and the indicator therefore shows middle position: $\alpha_m = 45°$.

The transmission cable consists of 3 wires. So temperature changes effect the line resistance R_L of each wire quite the same by ΔR_L. But as the denominator is equal to the numerator this effect is cancelled. α remains constant and is not a function of supply voltage or temperature T:

$$\alpha_m = \frac{U}{R_1 + 2\,(R_L + \Delta R_L)} \cdot \frac{R_1 + 2(R_L + \Delta R_L)}{U} = \text{const.}$$

$$\alpha_m \neq f\,[U,\, \Delta R_L(T)]$$

3 Measurements with Moving Coil Instruments

As shown in chapter 2, moving coil instruments are sensitive to the current passing through the coil. So far, a DC current I has only been suggested. It produces the deflection α of the pointer due to the scale equation. α is directly related to I:

$$\alpha = \frac{BNhd_c}{C} \cdot I$$

3.1 AC-MEASUREMENTS

If I is a function of time $I(t) = i$, the indication α changes correspondingly, in case the frequency of i is low enough. But in most technical applications the change of the measurand is faster than the indicator can follow. The torque inertia θ of the moving parts does not allow a fast response of the reading. For pure AC signals the pointer deflection becomes zero if their frequencies are high enough. 2002

Magnitude of the Mean Value of a Sine Current

Ex. In an experiment three different kinds of currents may be applied to an *MC* ammeter (Fig. 3.1-1). Firstly, a **DC-Current** I_- passes the coil. The

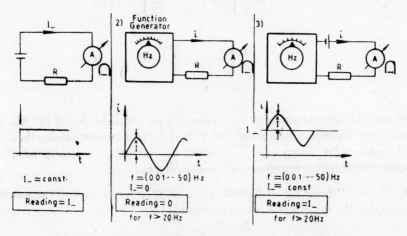

Fig. 3.1-1. Ammeter-reading for Different Types of Currents.

reading will be exactly this current. Secondly, a pure alternating current may flow. If its frequency is very low the indicator follows the current changes. But for increasing frequency, the indication change gets faster and the amplitude of the reading becomes smaller. Eventually the deflection will be zero. If thirdly, the DC and the AC are superimposed the display indicates only the DC component. The instrument obviously omits the AC for high frequencies. It shows the mean value of i:

$$\bar{i} = \frac{1}{T} \int_0^T i \, dt$$

In fact the torque inertia effects an integration of i. For a pure alternating current $i = \hat{I} \sin (\omega t)$, the mean value $\bar{i}$ will indeed be zero:

$$\bar{i} = \frac{1}{T} \int_0^T \hat{I} \sin \omega t \, dt$$

Substitution $\omega t = z$

limits $t = 0: \quad z = 0$

$$t = T: \quad z = \omega t = 2\pi$$

$$\bar{i} = \frac{\hat{I}}{T} \int_0^{2\pi} \sin z \, \frac{dz}{\omega} = \frac{\hat{I}}{2\pi} \int_0^{2\pi} \sin z \, dz$$

$$= \frac{\hat{I}}{2\pi} \Big[-\cos z \Big]_0^{2\pi} = 0 \qquad \text{q.e.d.}$$

For pure AC signals no direct indication can be obtained. But if the AC is transmitted through a bridge rectifier, (Fig. 3.1-2) the current passing the coil will contain a DC component $|\bar{i}|$ to which the instrument will respond. The bridge produces the magnitude $|i|$ of the current i.

Ex. For $i = \hat{I} \sin \omega t$ the indicated value is of interest. Due to the torque inertia it is the mean value of the magnitude $|\bar{i}|$

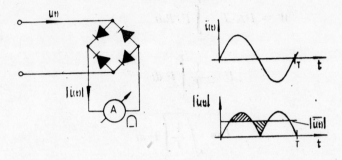

Fig. 3.1-2. **Moving-Coil Instrument with Bridge Rectifier.**

$$|\bar{i}| = \frac{1}{T} \int_0^T |i| \, dt = \frac{1}{T} \int_0^T \hat{I} \, |\sin \omega t| \, dt = \frac{2}{T} \hat{I} \int_0^{T/2} \sin \omega t \, dt$$

Substitution $\qquad \omega t = z$

$$dt = \frac{dz}{\omega}$$

Limits $\qquad\qquad t = 0: \qquad z = 0$

$$t = \frac{T}{2} : \qquad z = \frac{2\pi}{T} \frac{T}{2} = \pi$$

$$|\bar{i}| = \frac{2}{T} \hat{I} \int_0^\pi \sin z \, \frac{dz}{\omega} \quad = \frac{\hat{I}}{\pi} \left[-\cos z \right]_0^\pi$$

$$= \frac{2}{\pi} \hat{I} = 0.637 \, \hat{I}.$$

This is the value, the ammeter with a bridge rectifier, will respond to.

Root Mean Square Value (RMS) of a Sine Current

The magnitude of the mean value is usually not of interest. Instead the root mean square value (RMS) is needed as it is the value that can effect equivalent work (of a resistance R) just like a direct current I of the same amount during a certain time T:

$$W = I^2 R \cdot T = \int_0^T i^2 \cdot R dt$$

$$I^2 = \frac{1}{T} \int_0^T i^2 \, dt$$

$$I = \sqrt{\frac{1}{T} \int_0^T i^2 \, dt}$$

Ex. This is the definition of the RMS value, also called the effective value. For a current $i = \hat{I} \sin \omega t$, it may be calculated:

$$I^2 = \frac{1}{T} \int_0^T \hat{I}^2 \sin^2 \omega t dt \quad \sin^2 x = \frac{1}{2} - \frac{1}{2} \cos 2x$$

$$= 4 \frac{\hat{I}^2}{T} \int_0^{T/4} \left(\frac{1}{2} - \frac{1}{2} \cos 2\omega t \right) dt$$

Substituting $\qquad 2\omega t = z$

$$dt = \frac{dz}{2\omega} \quad \text{Limits} \ t = 0 : z = 0$$

$$t = \frac{T}{4} \ : z = 2 \frac{2\pi}{T} \cdot \frac{T}{4} = \pi$$

$$I^2 = 4 \frac{\hat{I}^2}{T} \left[\frac{1}{2} \frac{T}{4} - \frac{1}{2} \int_0^\pi \cos z \frac{dz}{2\omega} \right]$$

$$= 4 \frac{\hat{I}^2}{T} \left[\frac{T}{8} - \frac{1}{2} \cdot \frac{T}{2 \cdot 2\pi} \underbrace{\sin z \Big|_0^\pi}_{0} \right]$$

$$I^2 = \frac{\hat{I}}{2}$$

$$I = \frac{\hat{I}}{\sqrt{2}}$$

This current is usually the one of interest, which need be indicated.

Form Factor, AC- and DC-Scale

Moving coil instruments usually read the RMS value I because of its importance. But actually they are sensitive to the magnitude of the mean value $|\bar{i}|$. The RMS indication is, therefore, effected by calibrating the scale in RMS values, through the calibration factor F.

The RMS current I is different from the magnitude of the mean value $|\bar{i}|$ by the form factor

$$F = \frac{I}{|\bar{i}|}$$

As sinusoidal currents are the most important, the calibration is done for them and the sine form factor becomes

$$F_{\sim} = \frac{\hat{I}}{\sqrt{2}} \cdot \frac{\pi}{2\hat{I}} = 1.11$$

The instrument senses $|\bar{i}|$ from which I may be obtained as

$$I = 1.11 \times |\bar{i}|$$

The multiplication by the factor of 1.11 can be effected by using an additional scale for sinusoidal ACs, (Fig. 3.1-3 a), or employing an additional scaling resistor R_s for an instrument with one scale only. Modern ammeters, (fig. 3.1-3 b), employ a parallel scaling resistor R_s. It is switched on in the DC range but switched off for the AC range. Its value is arranged in

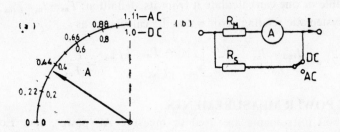

Fig. 3.1-3 (a) DC- and AC (sine)-Scale of a Moving Coil Instrument with Bridge Rectifier.

(b) Scaling Circuit for an Instrument with one scale only.

such a way that the current through the coil increases by a factor of 1.11 for the AC range. Thus the RMS value I can be read directly—Voltmeters have a series scaling resistor being shorted for the AC range.

Important Side note:

If waveshapes other than sinusoidal currents are applied still a multiplication of $|\tilde{i}|$ by 1.11 takes places as this constant always remains the same. It is entirely a feature of the instrument. But the magnitude of the mean value $|\bar{i}|$ is different. So the RMS value is certainly not equal to $1.11 \times |\bar{i}|$ for non sinusoidal measurands and the reading obtained does not give the RMS value. Another correction factor needs to be considered, which may be obtained in the same way as it was for sinusoidal currents. (If the indicated value I_{ind} should be calculated in this way, the magnitude of the mean value for the waveshape under suggestion needs to be determined first.)

Indicated current I_{ind} may be used to find the true RMS value I_{ns} for non sinusoidal waveshapes. For this purpose the form-factor F_{ns} is needed.

$$I_{ns} = F_{ns} \cdot |\bar{i}|_{ns}.$$

As per design of instrument,

$$I_{ind} = 1.11 \cdot |\bar{i}|_{ns}$$

Common for both equations is $|\bar{i}|_{ns}$ which allows to combine them and to solve for the true *RMS* current.

$$I_{ns} = \frac{F_{ns}}{1.11} I_{ind}$$

So the reading I_{ind} allows to determine the unknown RMS value of the non sinusoidal current I_{ns}. The form-factor F_{ns} is needed. It may be taken from a table or one can calculate it from its definition: $F_{ns} = I_{ns}/|\bar{i}|_{ns}$.

The systematic reading error ε may be calculated as

$$\varepsilon = \frac{I_{ind} - I_{ns}}{I_{ns}} = \frac{F_{\sim} \cdot |\bar{i}|_{ns} - F_{ns} \cdot |\bar{i}|_{ns}}{F_{ns} \cdot |\bar{i}|_{ns}} = \frac{F_{\sim}}{F_{ns}} - 1.$$

3.2 AC POWER MEASUREMENTS

Moving coil instruments are used to measure AC power, too. For this purpose the permanent magnet is replaced by an electro magnet. Details will be given in chapter 3.2.2. But first the definitions used must be made clear.

3.2.1 Definitions

We will care for the definitions of instantaneous power, active, reactive power, and power factor.

Instantaneous Power

The instantaneous power consumed by a load impedance Z is the product of instantaneous current i and voltage u of the load. Let us assume a sinusoidal voltage $u = \hat{U} \cdot \sin \omega t$. As Z may be of complex nature the current

$i = \hat{I} \sin(\omega t + \varphi)$ is shifted by the phase angle φ. The question for the components of the power p arises.

$$p = u \cdot i = \hat{U}\,\hat{I} \sin \omega t \cdot \sin(\omega t + \varphi)^*$$

$$= \hat{U}\,\hat{I}\,(\sin^2 \omega t \cos \varphi + \cos \omega t \sin \omega t \sin \varphi)^{**}$$

$$= \hat{U}\,\hat{I}\left[\frac{1}{2}(1 - \cos 2\omega t)\cos\varphi + \frac{1}{2}\sin 2\omega t \sin\varphi\right]$$

$$= \frac{\hat{U}\,\hat{I}}{2}\cos\varphi + \frac{\hat{U}\,\hat{I}}{2}\left[-\cos 2\omega t \cos\varphi + \sin 2\omega t \sin \varphi\right]^{***}$$

$$= \frac{\hat{U}\,\hat{I}}{2}\cos\varphi - \frac{\hat{U}\,\hat{I}}{2}\cos(2\omega t + \varphi) = p_- + p_\sim$$

The instantaneous power P consists of a constant component p_- and a purely alternating one $p_\sim$ (Fig. 3.2.1-1). The alternating component oscillates with double the frequency of voltage and current.

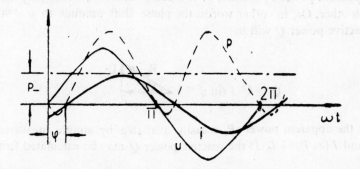

Fig. 3.2.1-1. Instantaneous values of Voltage u, Current i, and Power p of a Complex Load and Its Active Power p_-.

By forming the mean value p of $\bar{p}$, its alternating part $p_\sim$ does not effect $\bar{p}$ (as shown in chapter 3.1). Only the constant component p_- needs to be considered:

(*) $\sin(a + b) = \sin a \cos b + \cos a \sin b$

(**) $\sin^2 a = \frac{1}{2}(1 - \cos 2a)$ and $\cos a \cdot \sin a = \frac{1}{2}\sin 2a$

(***) $\cos a \cos b - \sin a \sin b = \cos(a + b)$

$$\bar{p} = \frac{1}{T} \int\limits_0^T p \; dt = \frac{1}{T} \int\limits_0^T p_- \; dt + \underbrace{\frac{1}{T} \int\limits_0^T p_\sim \; dt}_{0}$$

$$\bar{p} = p_-$$

Active Power

The active power P is equal to the constant power p_- contained in the instantaneous power p

$$P = p_- \frac{\hat{U}\hat{I}}{2} \cos \varphi = \frac{\hat{U}}{\sqrt{2}} \cdot \frac{\hat{I}}{\sqrt{2}} \cos \varphi$$

$$P = U \cdot I \cos \varphi$$

P is the product of the RMS values of u and i times the cos φ of the load. An instrument which can form the mean value $\bar{P}$ will indicate the active power, see chapter 3.2.2.

Reactive Power

The apparent power P_s consists of two components which are perpendicular to each other. Or, in other words, the phase shift amount to $\varphi = 90°$. So the reactive power Q will be

$$Q = U \cdot I \sin \varphi$$

As the apparent power P_s is easily available by simple measurements of U and I (as $P_s = U \cdot I$) the reactive power Q may be calculated from

$$Q = \sqrt{P_s^2 - P^2}$$

Power Factor

The ratio of active power P and apparent power P_s is called power factor λ. For sinusoidal quantities λ is equal to cos φ.

$$\lambda = \frac{P}{P_s}$$

$$\lambda_\sim = \frac{UI \cos \varphi}{UI} = \cos\varphi.$$

3.2.2   Single Phase Dynamometer

The most frequent method to measure electrical power employs a specially designed moving coil instrument, the dynamometer. It responds to the product of two currents. If one of them is the current through a load and if the other is proportional to the voltage across the load the indication will be proportional to the power consumed. But the dynamometer may also be used for voltage or current measurements see chapter 3.2.4.

Action Principle and Scale Equation

Dynamometers consist of a fixed field coil wound with heavy wire and a moving coil wound with light wire windings, (Fig. 3.2.2–1). Two spiral

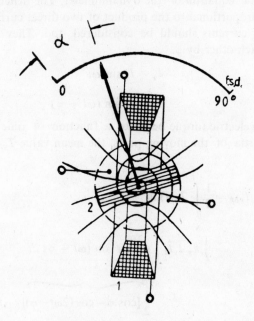

Fig. 3.2.2–1. Dynamometer.

springs provide the balancing counter torque. Coil 1 is shaped and positioned such that its field lines cross over the moving circle perpendicularly and produce the same flux density there. Coil 2 moves within a "homogeneous" field. Its induction B is proportional to the current I_1:

$$B_1 = k_1 \, N_1 \, I_1$$

The current I_2 flowing through coil 2, produces a force F on either side of coil 2:

$$F = N_2 h (B_1 \cdot I_2) = \overbrace{k_1 N_1 \cdot h N_2}^{k_2} \cdot I_1 I_2$$

The height h of coil 2 is the active length of the windings to produce this force. The coil radius $d_c/2$ forms the lever length due to which the electrical torque T_e causes the coil to move, $T_e = F\,2\,d_c/2$. With $k_3 = k_2.d_c$ we obtain T_e:

$$T_e = k_3\,I_1 I_2$$

The deflection α effects a mechanical counter torque $T_m = C \cdot \alpha$ which finally balances T_e:

$$k_3\,I_1 I_2 = C\alpha$$

$$k\,I_1 I_2 = \alpha$$

This is the scale equation of the dynamometer. The deflection angle α of the pointer is proportional to the product of two direct currents I_1 and I_2.

Alternating currents should be considered, too. They may be phase shifted from each other by φ.

$$I_1 = \hat{I}_1 \sin \omega t$$

$$I_2 = \hat{I}_2 \sin (\omega t + \varphi)$$

In this case the electric torque becomes a function of time $T_e(t)$. Due to the torque inertia of the moving parts, the mean value T_{eM} will effect the deflection:

$$T_{eM} = \frac{1}{T}\int_0^T T_e(t)\,dt$$

$$= \frac{1}{T} k_3\,\hat{I}_1 \hat{I}_2 \int_0^T \underbrace{\sin \omega t \cdot \sin (\omega t + \varphi)}_{*}\,dt$$

$$\underbrace{\tfrac{1}{2}[\cos \varphi - \cos (2\omega t + \varphi)]}$$

$$= k_3\,\frac{\hat{I}_1 \hat{I}_2}{2}\left[\frac{1}{T}\int_0^T \cos \varphi\,dt - \frac{1}{T}\int_0^T \cos (2\omega t + \varphi)\,dt\right]$$

As the mean value of a pure circular function equals zero (see chapter 3.1), the last integral need not be considered.

$$T_{eM} = k_3\,\frac{\hat{I}_1}{\sqrt{2}} \cdot \frac{\hat{I}_2}{\sqrt{2}} \cdot \cos \varphi = k_3\,I_1 I_2 \cos \varphi$$

The indication angle α is obtained from the equilibrium of T_{eM} towards the mechanical torque $T_m = C \cdot \alpha$.

$$k\,I_1 I_2 \cos \varphi = \alpha \quad \text{with } k = \frac{k_3}{C}$$

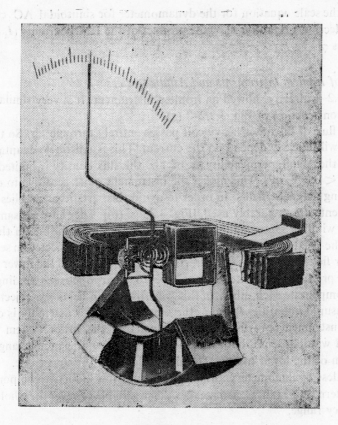

(a)

(b)

Fig. 3.2.2–2. (a) Ironless Electrodynamometer as Panel Instrument
(b) Astatic Type.

Hartmann and Braun, Germany

This is the scale equation for the dynamometer for sinusoidal **AC** currents. The deflection α is produced by the product of the RMS currents ($I_1 \cdot I_2$) and their cos φ.

Design of Ironless Instruments and Astatic Devices

Fig. 3.2.2–1 already showed an ironless instrument. It is very similar to the actual construction of Fig. 3.2.2–2 (a).

The flux of the outer heavy coil passes entirely through air. So the flux-density will linearly depend on the current. This is a distinct advantage. But due to the small permeability μ of air the flux density is limited to low values (< 0.1 V_s/m^2). This makes the instrument quite sensitive to external disturbing magnetic fields. In order to avoid this, astatic assemblies of two instruments are preferably used, [Fig. 3.2.2–2 (b)]. They have the same shaft. But the winding sense of the upper coils is opposite to the one of the lower coils. The fixed coils as well as the moving coils are connected in series. As an outer field shows the same direction for the upper and the lower moving coils it produces opposite torques due to their opposing winding sense. They compensate each other. In this way disturbing fields are rejected. For the measured quantities I_1 and I_2, the sensitivity of this assembly is doubled as two instruments are used. As both coils of the lower system have a reversed winding sense, both fields are reversed, resulting in no change of the direction of the lower torque. The shaft adds it to the upper one.

Ironless dynamometers normally have quite low inductive components of their internal resistances. This makes them suitable to cover the whole audio frequency range.

Design of Iron Closed Instruments

Quite often the sensitivity of ironless instruments is not sufficient. It may be considerably improved by leading the flux of the fixed coil mainly through an iron path, (Fig. 3 2.2–3). Only a small air gap is left inside which the moving coil is positioned. In this way the reluctance of the magnetic path is reduced essentially resulting in a far higher flux density within the air gap (> 0.5 V_s/m^2).

The need for an astatic assembly does not arise. The outer iron path reduces the impact of disturbing outer fields, thus shielding the air gap.

Connection of Voltage and Current Path

The outer coil usually serves to sense the current I of a load Z, (Fig 3.2.2–4). The internal resistances of the current path is therefore low. The moving coil measures the voltage U across Z. So its internal resistance is high. The current which passes through it is proportional to the measured voltage U. An additional resistor within the voltage path may be employed to match the instrument to different voltages, i.e. 220 V and 380 V (or 240 V and 415 V).

Fig. 3.2.2–3   Iron Closed Electrodynamometer as Panel Instrument.

Hartmann and Braun, Germany

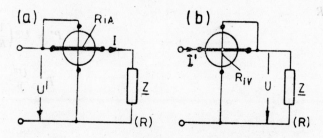

Fig. 3.2.2–4.   Connection of a Power Meter
(a)   for Sensing the True Current I
(b)   for Sensing the True Voltage U.

Active Power Measurement

The scale equation for the power meter is easily obtained. The previous current $I_1 = I$ which is the load current and the current $I_2 = U/R_{iV}$. U is the supply voltage of the load. R_{iV} is the internal resistance of the voltage path.

$$\frac{k}{R_{iV}} I \cdot U \cos \varphi = \alpha$$

$$I \cdot U \cos \varphi = c_w \alpha$$

The deflection α of the pointer is proportional to the active power.

$$R_{iV}/k = c_w$$

is the wattmeter constant. Its dimension is $[c_w] = 1 \text{ W/Scd}$.

As current and voltage cannot be measured exactly at the same time either the true current connection or the true voltage connection may be used, (Fig. 3.2.2-4). Whether the first or the second circuit will be used depends on the load Z with respect to the internal resistances of current and voltage path. For high $\underline{Z}$ the true current circuit avoids a high systematic measurement error. Due to a low load current I the voltage drop across the current path will be negligible. It hardly effects the voltage measurement. For low $\underline{Z}$ the true voltage circuit ensures a low error. The current through the voltage path does not effect the current measurement considerably.

The decision whether the connection of the instrument should be chosen to measure the current or the voltage truely, depends on the load resistance R with respect to the internal resistances R_{iA} and R_{iV} of the current and the voltage paths.

(a) *True current connection*

$$\Delta P_A = P' - P \quad P'$$

$$= U'I = I^2 (R + R_{iA})$$

$$P = I^2 R$$

(b) *True voltage connection*

$$\Delta P_V = P' - P \quad P' = UI^1$$

$$I' = \frac{U}{R} + \frac{U}{R_{iV}}$$

$$P' = U^2 \left(\frac{1}{R} + \frac{1}{R_{iV}}\right)$$

$$P = \frac{U^2}{R}$$

$$\Delta P_A = I^2\, R_{iA}$$

$$\Delta P_V = U^2\left(\frac{R + R_{iV}}{R \cdot R_{iV}} - \frac{1}{R}\right)$$

$$= \frac{U^2}{R_{iV}}$$

$$\varepsilon_A = \frac{\Delta P_A}{P} = \frac{R_{iA}}{R}$$

$$\varepsilon_V = \frac{\Delta P_V}{P} = \frac{R}{R_{iV}}$$

For large R the relative systematic error ε_A will be small.

For small R the relative systematic error ε_V will be small.

Equating $\varepsilon_A = \varepsilon_V$ allows to determine $R = R_D$, compared to which the decision for the choice of the circuit can be made: If $R > R_D$, then circuit (a) is correct. For $R < R_D$, circuit (b) should be chosen.

$$\varepsilon_A = \varepsilon_V$$

$$\frac{R_{iA}}{R_D} = \frac{R_D}{R_{iV}}$$

$$R_D = \sqrt{R_{iA} \cdot R_{iV}}$$

Side note: The same question for the proper connection arises if a volt- and an ammeter are used to determine an unknown resistance R_x. Following the same procedure as shown for the wattmeter connection another decision value R_D^* can be calculated as

$$R_D^* = \frac{R_{iA}}{2}\ {}^+_{(-)}\sqrt{\frac{R_{iA}^2}{4} + R_{iA} \cdot R_{iV}}$$

Reactive Power Measurement

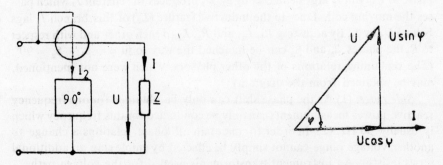

Fig. 3.2.2-5 Reactive Power Meter in a Single Phase Line.

The definitions of active (P) and reactive (Q) Power are

$$P = UI \cos \varphi \quad \text{and} \quad Q = UI \sin \varphi$$

They are phase shifted from each other by an angle of $\varphi = 90°$. A power-meter (as previously described) responds to the active power. But it can easily be made sensitive to the reactive power instead by shifting the current I_2 through the voltage path. The shift needs to be exactly $90°$ compared to the non-shifted case (as for Fig. 3.2.2-4). An auxiliary network tackles the problem, see Fig. 3.2.2-5. If we consider U as the reference, the voltage path will take into account the voltage component $U \sin \varphi$.

The HUMMEL circuit effects the necessary phase shift (Fig. 3.2.2-6).

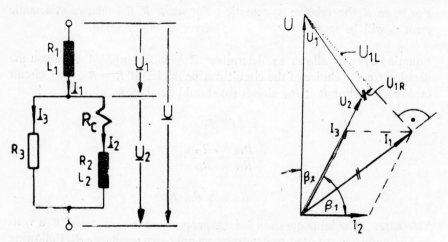

Fig. 3.2.2-6.   Phase Shifting Circuit and Phasor Diagram for the Voltage Path of a Single Phase Reactive Power Meter.

R_c represents the moving coil. Its current I_2 needs to be shifted with respect to the voltage U by $\varphi = 90°$. This is effected by a resistance R_3 and the inductors 1 and 2. They provide an active (R_1 and R_2) and a reactive component (L_1 and L_2) each. The total voltage U is composed from the leading voltage U_1 and the lagging one U_2. I_3 passes a resistor R_3. So I_3 is in phase to U_2. But I_3 lags behind U by β_2. U_2 produces the current I_2 which passes the moving coil. Due to the inductive feature (L_2) of this branch I_2 lags behind U_2 by β_1. By adjusting R_1, L_1, and R_2, L_2 to each other and with respect to R_3 the phases β_1 and β_2 can be matched the way to fit $\varphi = \beta_1 + \beta_2 = 90°$. (The remaining relations of the other phasors, which were not mentioned, may be obtained from the diagram).

Side notes: (1) As the phase shift can only be obtained for one frequency reactive power measurements can only be conducted for this frequency which is usually 50 Hz. (2) In order to maintain all phase relations a change to another voltage range cannot simply be effected by employing an additional series resistor. An instrument transformer is needed for the voltage path.

3.2.3 Electronic Power Measurement Device

The previously described instruments suffer from a limitation that they produce only a pointer deflection. But for automatic control devices, an electrical output signal is needed. It may be obtained from an electronic circuit. (Fig. 3.2.3-1) which also forms the product of voltage and current.

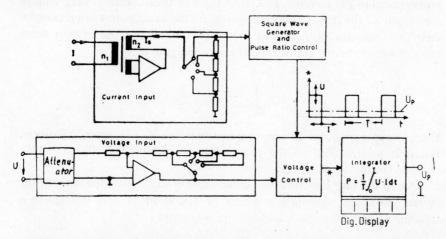

Fig. 3.2.3-1. Action Principle of an Electronic Power Measurement Instrument, (SIEMENS Power Meter B 4301)

The current input contains an active converter, employing an instrument transformer for the current. The secondary ampere windings are sensed (lower coil) and at the same time compensated by the current I_s through the upper coil. So there is nearly no flux inside the core of the current transformer. This means that nearly no inductance would effect the measured current I. As the primary side consists of a few heavy copper windings only, the voltage drop across the current input will be negligible. The input impedance of the current path will be therefore very low. The matching to different currents is effected by range resistors which are activated by a selector switch. As I_s is generated from many windings n_2 this current is low ($n_2 I_s = n_1 I$) and can easily be switched over by mechanical contacts. The measured current I needs not be interrupted when switching over to another measurement range. The circuit of the current path is extremely safe against overloading. In this case the operational amplifier will just go into saturation.

The voltage U to be measured is usually scaled down with the help of an input attenuator so that it can be processed by an operational amplifier which provides a high input impedance. Thus a loading effect of the measurand is avoided. The voltage range is selected within the feed back loop of the amplifier. A mechanical switch adjusts the output voltage to be below the amplifier saturation. Of course, the currents to be switched are low. But more important is the low voltage across the contacts instead of the measured U, which is usually high.

A time division multiplier (also called mark space multiplier) forms the product of $U \cdot I$. A block pulse generator of far higher frequency than that of the line is employed. The voltage U is rectified first. Another voltage proportional to I is rectified, too. If changes in them occur U may change the height of the pulses and I the width. As the block pulses are rectangles their area changes proportional to the product of $U \cdot I$ and so their mean value P will be, that P is obtained from an integrator:

$$P = \frac{1}{T} \int_0^T U \cdot I \, dt = U \cdot I$$

The integrator produces an output quantity which is, of course, a voltage. But it is proportional to the power P to be measured. A digital display presents the measured power.

Due to the very low inductive component of the input impedance of the current path, the described principle of operation of this instrument is sufficient to cover the whole audio frequency range.

3.2.4 Dynomometer for Voltage and Current Measurements

Voltage and current path may be connected in parallel to make up an ammeter. Or they are connected in series to make a voltmeter. Both instruments can cover the frequency range up to 20 kHz., if the ironless instrument type is employed.

Ammeter

Fig. 3.2.4-1 shows the ammeter circuit. The pointer deflection α will be

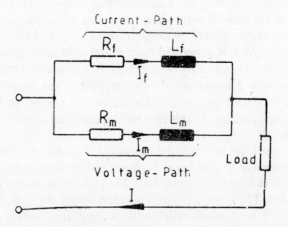

Fig. 3.2.4-1.   Dynamometer as Ammeter.

proportional to the product of the current through the fixed coil I_f and that through the moving coil I_m:

$$\alpha = k \cdot I_f \cdot I_m$$

I_f will be one part of the measured current I and I_m the other.

$$I_f = aI$$

$$I_m = bI \quad \text{with} \quad a + b = 1$$

So α becomes

$$\alpha = c_I \; I^2$$

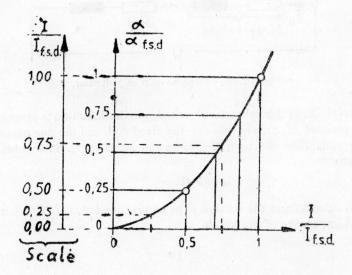

Fig. 3.2.4-2.   Non-Linear Current Scale of Dynamometer.

This is the scale equation. The instrument constant $c_I = k \cdot ab$. If I is an alternating current $I(t)$, the currents I_f and I_m are alternating as well. They alternate with time $I_f(t)$ and $I_m(t)$ and they are in phase with each other. Whether they might be of block shape or of sinusoidal or other shape is not important. Due to the torque inertia of the moving coil an integration of $I^2(t)$ is effected. The squared mean value $\overline{I^2}$ of the squared measured quantity $I^2(t)$ is formed which is the squared RMS value $(\overline{I^2})$. The pointer responds directly to $\overline{I^2}$.

$$\alpha = c_I \underbrace{\frac{1}{T} \int_0^T I^2(t) \, dt}_{\overline{I^2}}$$

As usual the RMS value I is of interest (instead of I^2) and the spacing of the scale divisions is not linear. If for instance full scale deflection (fsd) is reached for 1 A, only a quarter of fsd is produced for 0.5 A. The space width between the scale divisions increases for increasing deflection α, (Fig. 3.2.4-2).

Voltmeter

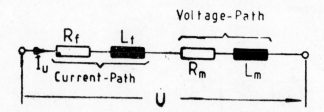

Fig. 3.2.4-3.   Dynamometer as Voltmeter.

Fig. 3.2.4-3 shows the voltmeter circuit. Again the deflection is proportional to the product of current passing the fixed coil and the one passing the moving coil. They are both the same and equal to I_u. So we obtain the indication

$$\alpha = k\, I_u^2$$

But I_u is caused from the voltage U to be measured: $I_u = U/Z$, where Z is the total impedance of the circuit.

$$\alpha = \frac{k}{Z^2} \cdot U^2$$

$$\alpha = c_u \cdot U^2$$

The scale equation shows that this voltmeter produces a deflection proportional to U^2. But as the RMS value U is usually of interest, the scale is not linear, see above.

3.2.5   Three Phase Power Measurements
In 3 phase lines, power measurements can always be effected by using three instruments. They are certainly necessary for complete lines (i.e. 4 wire system). But for incomplete lines (i.e. 3 wire system) two instruments will do.

Complete 3 Phase Line
The most common type of load consists of unbalanced consumers connected to the three phases R, Y, B with respect to neutral (N). One could consider the sum of phase currents $I_R + I_Y + I_B$ as going into the load and the

current I_N of the neutral wire as coming from it. Consequently it will be sufficient to measure the power of the three phases. If their RMS currents I_R, I_Y and I_B and the RMS voltages of the phases U_R, U_Y and U_B are considered, the total apparent power S will result from

$$S = I_R U_R + I_Y U_Y + I_B U_B$$

But normally ordinary power meters are used. They measure the active power and produce an indication α proportional to $I \cdot U \cdot \cos \varphi$. So the total active power will result from the sum of three readings, (Fig. 3.2.5-1) as

$$P = I_R U_R \cos \varphi_R + I_Y U_Y \cos \varphi_Y + I_B U_B \cos \varphi_B$$

$$= c_w (\alpha_R + \alpha_Y + \alpha_B)$$

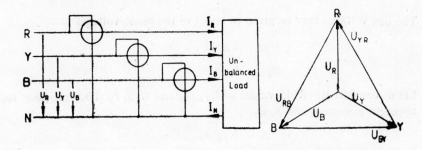

Fig. 3.2.5-1.   Three Phase Power Measurement for Unbalanced Load.

For the case of balanced load $I_R = I_Y = I_B = I$ and the phase angles are equal: $\varphi_R = \varphi_Y = \varphi_B = \varphi$. The phase voltages are alike, $U_R = U_Y = U_B = U$. So the active power measurement can be effected with the help of only one wattmeter. But its reading needs to be multiplied by 3:

$$P = 3 \cdot UI \cos \varphi$$

$$= 3 \cdot c_w \cdot \alpha$$

The neutral conductor carries no current.

Incomplete 3 Phase Line: Two Wattmeter Method
As balanced loads are quite common, the neutral conductor is usually not available. But still the power needs to be measured. This can be done using the two wattmeter method, (Fig. 3.2.5-2), which measures the power for unsymmetrical load as well.

Now I_R and I_B are considered as going into the load and I_Y as coming out. So two instruments will allow a total power determination.

The following definitions will be needed: The sum of all currents is zero, so

$$I_R + I_B = -I_Y.$$

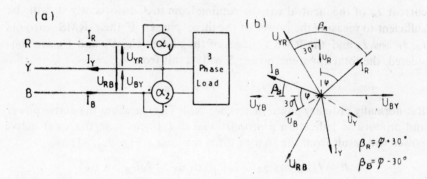

Fig. 3.2.5-2.   (a) Two Wattmeter Circuit (b) Phasor Diagram for Balanced Load.

The line voltages may be given in terms of the phase voltages:

$$U_{YR} = U_R - U_Y$$

$$U_{YB} = U_B - U_Y$$

Let us consider the RMS values of U_{YR}, I_R and U_{YB}, I_B. They produce the apparent powers S_1 and S_2 as

$$S_1 = U_{YR} \cdot I_R$$

and

$$S_2 = U_{YB} \cdot I_B$$

We form their sum $(S_1 + S_2)$ and prove what it will produce:

$$S_1 + S_2 = (U_R - U_Y) I_R + (U_B - U_Y) I_B$$

$$= U_R I_R + U_B I_B + U_Y \underbrace{(-I_R - I_B)}_{I_Y}$$

The comparison to the first equation in this chapter shows that the sum equals the total apparent power:

$$S = S_1 + S_2$$

Active Power Measurement P
What is true for S will be true for the active power P as well. We measure P_1 and P_2 so the total power P is

$$P = P_1 + P_2 = c_w (\alpha_1 + \alpha_2)$$

α_1 and α_2 are the readings in scale divisions of the two wattmeters. c_w is the wattmeter constant.

For the special case of balanced load this could be proved, see phasors of Fig. 3.2.5-2 (b).

$$P_1 = U_{YR} \cdot I_R \cos(\varphi + 30°)$$

$$P_2 = U_{YB} \cdot I_B \cos(\varphi - 30°)$$

The line voltages are equal $U_{YR} = U_{YB} = U_L$. The currents are equal, too:

$$I_R = I_B = I.$$

$$P = P_1 + P_2 = U_L\, I\, \underbrace{[\cos(\varphi + 30°) + \cos(\varphi - 30°)]}_{\underbrace{2\cos 30° \cos \varphi}_{\sqrt{3}}}$$

$$P_1 + P_2 = U_L\, I \sqrt{3} \cdot \cos \varphi$$

As $U_L = \sqrt{3} \cdot U$, we obtain indeed the total active power

$$P_1 + P_2 = 3UI \cos \varphi = P.$$

Two wattmeters are frequently mounted on the same shaft. In this way the sum of $(P_1 + P_2)$ is immediately effected and a direct reading of the total active power P is available.

Reactive Power Measurement Q

To determine Q the difference of the readings is investigated, but again for balanced load only:

$$P_2 - P_1 = U_L I \underbrace{[\cos(\varphi - 30°) - \cos(\varphi + 30°)]}_{\underbrace{2 \sin 30° \sin \varphi}_{1}}$$

$$= \sqrt{3} UI \cdot \sin \varphi.$$

But the total reactive power $Q = 3\,UI \cdot \sin \varphi$. So the readings need be multiplied by $\sqrt{3}$:

$$Q = \sqrt{3}\, c_w\, (\alpha_2 - \alpha_1).$$

By mounting two wattmeters on the same shaft a direct reading of the total reactive power is possible. But both instruments must act contrary to each other to produce a deflection proportional to $(P_2 - P_1)$. The factor of $\sqrt{3}$ should be a feature of the scale.

Side note: Reactive power of a one phase load (and its phase angle even, or $\cos \varphi$) can be measured directly if all three phases are available. In this case an ordinary watt meter may be used (or a cross coil dynamometer for $\cos \varphi$). Its current path should carry the load current. But the voltage path needs to be connected to a voltage which is shifted towards the phase voltage by

90° (see also chapter 3.2.2), for example U_{YB} if the load was connected to U_R. The reading needs to be divided by $\sqrt{3}$ to obtain Q, for U_{YB} is $\sqrt{3}$ times the actual voltage U_R.

Phase Angle φ

The phase angle φ of a balanced load may be easily determined. It is

$$\tan \varphi = \frac{Q}{P} \frac{\sqrt{3}\, c_w(\alpha_2 - \alpha_1)}{c_w(\alpha_1 + \alpha_2)}$$

$$\varphi = \operatorname{arc\,tan} \frac{\alpha_2 - \alpha_1}{\alpha_1 + \alpha_2} \sqrt{3}$$

(Mind the difference of $(\alpha_2 - \alpha_1)$. In case it is small, see error calculation). If a direct indication of φ or cos φ is needed a cross coil dynamometer may be advantageously used, see next chapter.

The two wattmeter method proves to be quite versatile. Honouring the inventor it is often referred to as ARON-circuit.

3.3 POWER FACTOR MEASUREMENT

The ARON circuit is one possibility to determine the power factor. But no direct reading is obtainble. Another restriction is that it can only be applied to a three phase system.

A further possibility, which also can work on a single phase line, results from the active power definition:

$$P = UI \cos \varphi \Rightarrow \cos \varphi = \frac{P}{UI}$$

The measurements of power P and of the RMS values of voltage U and load current I allow to calculate cos φ. But three measurement errors will add up in cos φ.

A cross coil dynamometer avoids these disadvantages and responds with a direct reading (Fig. 3.3-1). The fixed coil of the stator carries the load current I. It is of low impedance. The two moving coils are rigidly mounted on the shaft having an angle of 90° to each other. They provide a high impedance because they are used as voltage paths. There is no mechanical counter torque. T_1 may be obtained clockwise. It is controlled by T_2 which acts anticlockwise due to the winding sense of the related coil. The currents I_1 and I_2 are supplied by four tiny soft gold strips. The load current I generates a homogeneous magnetic field within the airgap. Together with I_1 and I_2 the torques T_1 and T_2 are produced. T_1 is proportional to the active power of the load but also proportional to the sine of the angular position α. The torque T_2 is phase shifted by 90° towards T_1 due to the HUMMEL-circuit. So T_2 will be proportional to the reactive power but also to the cosine of the coil position α:

$$T_1 \sim UI \cos \varphi \sin \alpha$$

$$T_2 \sim UI \sin \varphi \cos \alpha$$

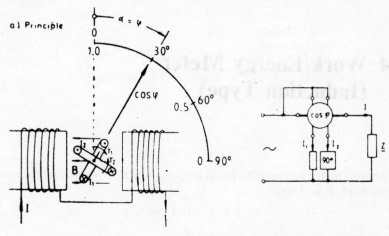

(b) **Circuit-Connection for** cos φ Measurement.

Fig. 3.3-1.   Electrodynamic Quotient-Meter.

For the indication position α of the pointer both torques are in equilibrium.

$$T_1 = T_2$$

$$UI \cos \varphi \sin \alpha = UI \sin \varphi \cos \alpha$$

$$\frac{\sin \alpha}{\cos \alpha} = \frac{\sin \varphi}{\cos \varphi}$$

$$\tan \alpha = \tan \varphi$$

$$\alpha = \varphi.$$

This is the scale equation of the cross coil moving coil instrument. The phase angle ϕ may directly be indicated. The pointer position α gives φ. But usually the power factor cos φ is needed. So a non-linear scale is employed. For an ohmic load cos $\varphi = 1$. Zero deflection will be indicated for this case.

4 Work/Energy Meter (Induction Type)

The workmeter responds to the active component of work/energy consumed by a load. It measures

$$W = P \cdot t = U \cdot I \cdot \cos \varphi \cdot t.$$

4.1 DESIGN OF INSTRUMENT AND GENERATION OF CURRENT AND VOLTAGE FLUX

The induction work meter (Fig. 4.1–1), consists firstly of a current coil (4), wound on the current iron (5). A few heavy windings carry the load current I which causes the flux ϕ_I. Due to a wide airgap which the flux ϕ_I has to pass the reluctance R_m of the magnetic path is high. Both makes the inductance $L_I = N^2/R_m$ small. Secondly there is the voltage coil (2) wound with a lot of windings. The load voltage U causes a current I_U through it which

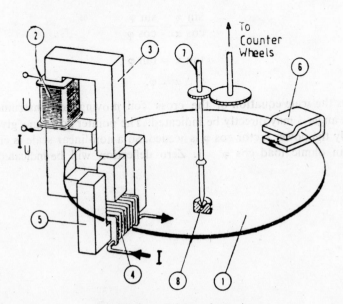

Fig. 4.1–1. (Origin unknown)

Action Principle of an Induction Type Work Meter.

is considerably phase shifted towards U. The coil current generates the voltage flux ϕ_U. It is nearly 90° phase shifted with respect to U especially because the reluctance R_m is low due to a small airgap. This results in a high inductance L_U of the voltage path.

4.2. EDDY CURRENTS WITHIN THE ARMATURE DISC

An aluminium disc acts as an eddy current drive. The fluxes ϕ_I and ϕ_U generate eddy currents I_{EI} and I_{EU}. They in return interact with ϕ_U and ϕ_I respectively and cause a circumference force which drives the disc. The revolutions of the disc are counted. A permanent magnet (6) acts as an eddy

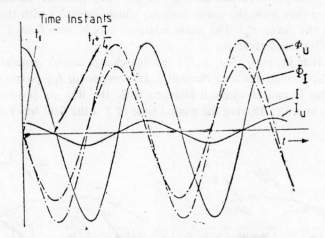

Fig. 4.2–1. Induction Type Work Meter. Currents and Fluxes of Voltage and Current Paths for Active Load.

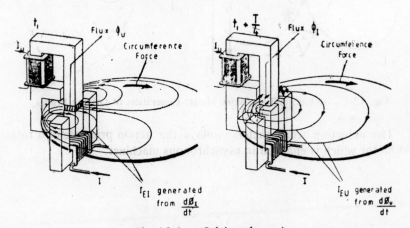

Fig. 4 2–2. (Origin unknown)

Induction Type Work Meter. **Voltage Flux** ϕ_U and Current Flux ϕ_I together with **Eddy Currents** I_{EI} and I_{EU} for $t_1 = 0$ and $t_2 = t_1 + T/4$.

current brake. The relations of I_U, ϕ_U, I, and ϕ_I to each other are depicted in Fig.4.2–1.

For the time instants $t_1 = 0$ and $t_2 = T/4$ the three-dimensional depictions of Fig 4.2–2 allow an easy understanding of the interactions.

4.3 DERIVATION OF THE ELECTRICAL DRIVING TORQUE

Fig. 4.3–1 depicts the instantaneous situation of the previous figures but now within the drawing plane. The current flux ϕ_I flows through the area elements 1*a* and 1*b*, and the voltage flux ϕ_U through element 2.

At the time instant t_1 the flux change $d\phi_I/dt$ generates eddy currents I_{E1a} and I_{E1b} around the area elements 1*a* and 1*b*. They pass the element 2. At this location they form the vector sum I_{EI} which interacts with the flux ϕ_U producing the force F_2. The phase relations may be seen from Fig. 4 2–1 and the following derivations.

At the time instant $t_2 = t_1 + T/4$ the flux change $d\phi_U/dt$ generates eddy currents I_{E2} around the area element 2. Its component I_{EU} passes through the locations 1*a* and 1*b* where it interacts with the flux ϕ_I producing the forces F_{1a} and F_{1b}. During the passed time of $T/4$ the field has rotated by

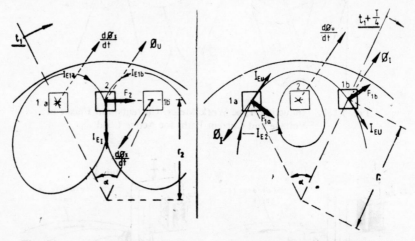

Fig. 4.3-1.   Induction Type Work Meter: Generation of Driving Forces.

$\alpha/2$. The induction work meter employs the action principle of a rotary field motor which is engaged in asynchronous machines.

ϕ_U and I_{EU}

Due to the considerable inductance L_U of the voltage coil, its current I_U will nearly lag by 90° behind the load voltage U. And so does the Flux ϕ_U.

$$U = \hat{U} \sin \omega t$$

$$I_U = \hat{I}_U \cos \omega t \quad \hat{I}_U = \frac{\hat{U}}{\omega L_U}$$

$$\phi_U = B_U \cdot A_U = \mu H_U \cdot A_U$$

$$= \mu A_U \frac{I_U N_U}{l_U}$$

$$= \underbrace{\frac{\mu A_U N_U}{l_U} \frac{\hat{U}}{\omega L_U}}_{\hat{\phi}_U = k_U \cdot \hat{U}} \cos \omega t$$

ϕ_I and I_{EI}

The current coil provides a very low inductance L_I. The current through the coil is the load current I. It lags behind U by φ due to the features of the load and produces the flux ϕ_I.

$$I = \hat{I} \sin (\omega t + \varphi)$$

$$\phi_I = B_I \, A_I = \mu H_I \cdot A_I$$

$$= \mu A_I \frac{I \, N_I}{l_I}$$

$$= \underbrace{\frac{\mu A_I \, N_I}{l_I} \cdot \hat{I}}_{\hat{\phi}_I = k_I \cdot \hat{I}} \sin (\omega t + \varphi)$$

The meanings of the symbols are selfevident from previous chapters.

Inside the disc are eddy currents I_{EU} generated due to the flux change $d\phi_U/dt$.

$$e_U = - \frac{d\phi_U}{dt} N \qquad N = 1$$

$$= k_U \hat{U} \omega \sin \omega t$$

$$I_{EU} = \frac{e_U}{R_{EU}} = \underbrace{\frac{k_U \omega \hat{U}}{R_{EU}}}_{\hat{I}_{EU}} \sin \omega t$$

R_{EU} is the disc resistance for the eddy currents I_{EU}.

The flux change $d\phi_I/dt$ generates eddy current I_{EI} inside the disc.

$$e_I = - \frac{d\phi_I}{dt} N \qquad N = 1$$

$$= - k_I \hat{I} \omega \cos (\omega t + \varphi)$$

$$I_{EI} = \frac{e_I}{R_E} = - \underbrace{\frac{k_I \omega \hat{I}}{R_{EI}}}_{\hat{I}_{EI} = k^* \cdot \hat{I}} \cos (\omega t + \varphi)$$

R_{EI} is the disc resistance for the eddy currents I_{EI}.

By interaction of ϕ_U and I_{EI} the force F_2 at area element 2 is obtained. It provides one driving force F_2 on the disc. It is calculated hereafter. Two other forces F_{1a} and F_{1b} are available from the elements $1a$ and $1b$ due to the interaction of ϕ_I and I_{EU}. They are perpendicular to I_{EU} and to ϕ_I. But only their components, which are perpendicular to the disc radius are responsible for the drive of the disc. They could also be derived following the same procedure as for F_2. Both, F_2 and F_{1a} as F_{1b}, are proportional to the active power of the load. But this is shown for F_2 only.

$$\mathbf{F_2} = k\,(\mathbf{I}_{EI} \times \Phi_U)$$

This equation is commonly valid. But $\mathbf{I}_{EI}$ and Φ_U are perpendicular to each other and so $\boldsymbol{F_2}$ becomes

$$F_2 = K I_{EI}\,\phi_U$$

$$= \underbrace{K k_U k^*\, \hat{I}\hat{U}}\ \ \underbrace{\cos(\omega t + \varphi)\,\cos\omega t}$$

$$= \hat{F}_2 \cos\omega t\ \overbrace{(\cos\omega t \cos\varphi - \sin\omega t \sin\varphi)}$$

$$= \hat{F}_2\,\underbrace{(\cos^2\omega t)}\,\cos\varphi - \underbrace{\cos\omega t \sin\omega t}\,\sin\varphi)$$

$$\overbrace{\tfrac{1}{2}(1 + \cos 2\omega t)}\qquad\qquad \overbrace{\tfrac{1}{2}\sin 2\omega t}$$

$$= \frac{\hat{F}_2}{2}\,(\cos\varphi + \cos\varphi \cos 2\omega t - \sin\varphi \sin 2\omega t)$$

$$= F_{2-} + F_{2\sim}$$

F_2 consists of a constant component and an alternating one. But only F_{2-} drives the disc. Due to its torque inertia the mean value of F_2 is applicable. It is

$$\bar{F}_2 = \frac{1}{T}\int_0^T F_2(t)\,dt = \frac{\hat{F}_2}{2}\cos\varphi$$

Looking back to the above equations we may additionally state that $\hat{I}_{EI}$ is certainly proportional to the maximum of the load current $\hat{I}$ and that $\hat{\phi}_U$ is proportional to the peak voltage $\hat{U}$ of the load. So F_2 can be written as:

$$F_2 = k_F \frac{\hat{I}}{\sqrt{2}} \frac{\hat{U}}{\sqrt{2}} \cos \varphi = k_F \overbrace{IU}^{P} \cos \varphi$$

and in this way it is proved that the circumference force is proportional to the active power P of the load.—The same result could be shown for F_{1a} and F_{1b}. Their perpendicular components F_p to the disc radius r_1, as well as F_2 together with the radius r_2 (see fig. 4.3-1) produce the actual driving torque T_e:

$$T_e = F_2 r_2 + 2F_p r_1 = k_T \cdot P$$

The driving torque T_e is balanced by the counter torque T_B of the brake magnet. T_B is produced by the reaction field of eddy currents as a response to the original field of the brake magnet (Fig. 4.3-2). Due to the induction

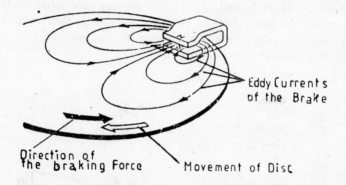

Fig. 4.3-2 (*Origin unknown*). Eddy Currents of the Brake Magnet.

law involved, the braking torque T_B will be proportional to the angular disc speed which can be expressed in terms of revolutions n per minute: $T_B = k_B n$. The resulting revolutions n are obtained from the equilibrium condition:

$$T_e = T_B$$

$$k_T \cdot P = k_B \cdot n$$

$$n = \frac{k_T}{k_B} P = kP$$

The disc revolutions per minute n are proportional to the active power P of the load.

4.4. COUNTER WHEELS AND INDICATION OF WORK

Fig. 4.1-1 shows that the shaft of the disc drives a counter which consists of indicator wheels with figures on them. This counter adds up all revolutions during the reading interval $(t_b - t_a)$ permanently. The active work W

is the active power P times the time passed since the last reading. As the counter effects a continuous integration, changing P (that means changing n) will be allowed, too. The reading is

$$z = z_b - z_a = \int_{t_a}^{t_b} n\, dt = k \int_{t_a}^{t_b} P\, dt = k \cdot W$$

The meter constant k is usually given on the front side of the instrument ranging from 600 to 4000 rev./kWh for single phase instruments. Three phase devices usually provide $k = 150$ rev./kWh.

4.5   ERRORS AND THEIR COMPENSATION

The correct indication of the active work can be effected only if the phase shift between the fluxes ϕ_U and ϕ_I is exactly 90° (for pure active load). The theory proves this need. But in practice the voltage path is not a pure inductance and the current path is not a pure ohmic resistance as it should be. So measurement errors arise.

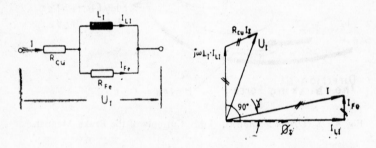

Fig. 4.5-1.   Current Path: Equivalent Circuit and Phasor Diagram.

Current Drive

Fig. 4.5-1 shows the equivalent circuit of the current path and its related phasor diagram which shows that the load current I is unfortunately not in phase with ϕ_I. ϕ_I lags slightly behind I by the error angle γ (exaggeratedly depicted). This is because the current path may be considered as an iron choke with an airgap. Magnetizing the choke needs the current I_{LI}. The iron losses are covered by I_{Fe}. Both I_{LI} and I_{Fe} are parts of $I = I_{LI} + I_{Fe}$.

The flux ϕ_I is actually in phase to I_{LI} and not to I which it should be. This effect is compensated with the help of a magnetic shunt path, which is part of the voltage path.

Voltage Drive

As depicted in Fig. 4.5-2 a magnetic shunt path, as part of the voltage iron, is shown. It serves, firstly, the purpose to achieve a high induction of the voltage coil in order to obtain a phase angle of 90° between the load voltage U and the driving flux ϕ_U of the voltage iron. Due to the iron losses and the copper resistance of the coil this could certainly not be reached for the total flux $\phi_{tot} = \phi_U + \phi_S$. But there is a shunt path which bypasses

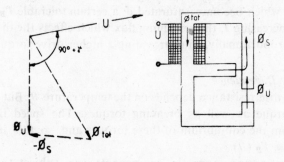

Fig. 4.5-2. Phasor Diagram of the **Voltage Path**
Showing the Compensation of Current
Phase Error.

the shunt flux ϕ_S. For the voltage drive only ϕ_U is used. For ϕ_U we can easily obtain a phase shift towards U, of 90° and even more, as it is actually done by the additional angle γ which was the error angle of the current path. ϕ_U lags behind U by 90° $+ \gamma$, thus fitting the need to have a phase shift between ϕ_U and ϕ_I of exactly 90°. In this way, the error angle γ of the current path, is compensated with the help of the voltage path which is shunted.

As it is difficult to adjust towards exact compensation, a shorted winding is mounted on the current iron. It acts as an additional load which may be easily changed by shifting a shorting blank to the proper position to fit all phase conditions (not depicted).

Friction and other Braking Torques

Both paths not only produce the driving torque as described but also imply small braking torques $T_{BU} = k_{BU} \cdot \phi_U{}^2 \cdot n$ and $T_{BI} = k_{BI} \cdot \phi_I{}^2 \cdot n$.

They are proportional to the RMS values of the related fluxes. It is easy, to reach compensation for the voltage path, for U is constant. So an additional adjustable iron piece is mounted near the airgap that splits the voltage flux ϕ_U to provide a slightly phase shifted additional voltage flux. This way an additional driving torque is effected to compensate T_{BU}. Actually it is made slightly greater than needed for this purpose in order to compensate the friction of the bearings, too. —The voltage drive is composed in such a way as to secure a guaranteed start of the disc motion for 5% of the nominal current I_N. But for no load at all, the work meter should not start running at least not until 20% of the nominal voltage U are exceeded.

The compensation of the braking torque T_{BI} of the current path causes problems as the load current I may change in wide ranges. But still, good results may be achieved by employing an additional magnetic shunt for the current iron which becomes saturated if a certain tolerable T_{BI} is exceeded. By further increasing I, the remaining flux which effects the disc drive, rises more than proportionally, thus providing a higher drive torque.

Temperature Dependence

The specific disc resistance depends on the temperature ϑ. But this concerns the drive torque as well as breaking torques. The speed of the disc is obtained from the equilibrium of these torques and so this influence cancels: $T_e(\vartheta) = T_B(\vartheta)$.

The electric and magnetic components are subject to temperature changes. But by appropriate combination of the materials used, a sufficient compensation is reached. For the brake magnet the magnetic path consists of two iron pieces. One provides a negative temperature coefficient, and the other a positive one, compensating each other.

Total Error Over Load Current

In spite of all these measures a certain instrument error remains. It can be measured as

$$\varepsilon\left(\frac{I}{I_N}\right) = \frac{W_{\text{ind}}\left(\frac{I}{I_N}\right) - W_{\text{true}}\left(\frac{I}{I_N}\right)}{W_{\text{true}}\left(\frac{I}{I_N}\right)}$$

I = load current $\qquad$ W_{ind} = indicated work
I_N = nominal meter current $\qquad$ W_{true} = true value of work.

It is agreed standard for work meters that its error ϵ does not exceed the calibration error ϵ_C for new instruments and eventually not ϵ_L for instruments after long term use (less than 20 years), specified as follows:

$$\left. \begin{aligned} |\epsilon_C| &\leqslant 3 + 0.05\,\frac{I}{I_N} + 0.05\left(1 + 0.1\,\frac{I}{I_N}\right)\tan\varphi \\ |\epsilon_L| &\leqslant 2\cdot|\epsilon_C| \end{aligned} \right\}$$

$$\text{for} \left\{ \begin{aligned} 0.05 &\leqslant \frac{I}{I_N} \leqslant 1.25 \\ 0.5\ &\leqslant \cos\varphi \leqslant 1 \end{aligned} \right.$$

These functions of ϵ_C and ϵ_L are depicted over I/I_N, in Fig. 4.5-3. Due to calibration regulations the instrument error should be negative only. This means a reading which is in favour of the consumer.

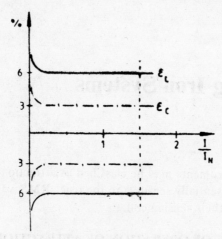

Fig. 4.5-3.   Permissible Limits of Calibration Error ϵ_C and the
Error ϵ_L after Long Term Use.

ϵ_C limits the calibration error of the instrument. The calibration is
effected with the help of the brake magnet which can be shifted by means of
a displacement mechanism. It allows to set the meter constant k to its
nominal value, i.e. $k = 650$ Rev./kwh.

5 Moving Iron Systems

Moving iron instruments may be classified as attraction type and repulsion type. Both are essentially sensitive to the true **RMS** value of the current passing through their sensing coil.

5.1 PRINCIPLE OF OPERATION OF ATTRACTION AND REPULSION TYPE

Attraction type
The moving iron system consists simply of a coil that attracts a soft iron piece. It is pulled into the stronger magnetic field inside the coil once a current I to be measured is applied. The pointer deflection is controlled by a spring, (Fig. 5.1–1).

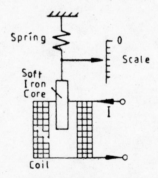

Fig. 5.1–1 Soft Iron System, Attraction Type.

This type is quite rarely used. Due to its construction, it is quite sensitive to outer vibrations and accelerations which restrict its sensitivity. But it is in use for voltage testers, just to distinguish the line voltage from the phase voltage, such as 380V from 220 V or 415 V from 240 V etc.

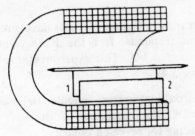

Fig. 5.1–2. Soft Iron System, Repulsion Type.

Repulsion Type

Two pieces 1 and 2 are mounted close to each other inside the coil (Fig. 5.1–2). If a current passes through the coil, the field produced magnetizes both pieces, and they will produce north poles on their one ends and south poles on their other ends. As like poles are adjacent to one another, a repulsion of piece 1 from piece 2 takes place. In order to enable a movement, piece 1 is mounted on the shaft and piece 2 is rigidly fixed to the coil. A spiral spring provides the control torque (not depicted).

5.2 CURRENT MEASUREMENTS WITH SOFT IRON INSTRUMENTS

Ex. Any soft iron instrument may be employed for the following experiments. A DC current should be applied, let $I = 1$ A. A certain deflection α will be noticed. Then this current may be reversed, but the same α will occur (Fig. 5.2–1). Even if the current alternates blockwise ($\hat{I} = I_-$) the reading α stays the same, independent from frequency and pulse ratio.

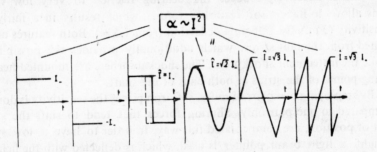

Fig. 5.2–1. Different Currents of the same RMS value I.

After that a sine current of $\hat{I} = I \cdot \sqrt{2} = 1.414$ A and 50 Hz should pass the instrument coil. The same indication α will be maintained. This is the case also for a changed frequency, for example 25 Hz of the sine. Applying a triangular current of $\hat{I} = I \cdot \sqrt{3} = 1.73$ A again causes the same deflection α even for different frequencies. As always the same indication α is reached, it is obvious that all these different currents have something in common to which the instrument responds. It is the RMS value, which is indeed the same for all currents: $I = 1$ A. This experimental result can also be proved theoretically (chapter 5.4).

Side Note: Moving iron Instruments should be used preferably for AC measurements. For DC the remanence of the soft iron pieces causes a slightly different reading for reversed polarity.

5.3 DESIGN OF ACTUAL MOVING IRON INSTRUMENTS

For actual measurements, only repulsion instruments are in use. The concentric vane type as shown in Fig. 5.3-1 and 5.3-2 is employed for panel meters. The radial vane type of Fig. 5.3-3 may serve for very sensitive precision instruments. All of them have no current carrying components being parts of the moving system. The spring provides the control torque only. An air damper as in Fig. 5.3-1 or an eddy current damper as in 5.3-2 ensure a short settling time. The construction allows only a rotary movement and for instrument 5.3-2 the deflection may be even 250°. Due to a certain shape of the concentric vanes a linearization of the scale is effected, see chapter 5.4. From certain deflections α onwards the iron vanes get partly saturated which changes the inductivity and makes it a function of α. The coil is specially wound to obtain a shape that aims for linearization, too. Due to the voluminous coil, the system enjoys good cooling which allows high over load currents for long duration without the danger of overheating.

Moving iron instruments are not very sensitive compared to moving coil instruments. But the construction in Fig. 5.3-3 employs a taut strip suspension which effectively reduces the bearing friction to very low values. This allows to have a soft restoring torque, which results in a fairly high sensitivity (3 mA for full scale deflection from 7.5 V). Both features are obtained from the taut strips, which additionally provides the possibility to make the instrument shockproof. The shockabsorbers are mounted near the fixing points of the strips at both ends of the shaft.

By employing two pairs of radial soft iron pieces the sensitivity is doubled, compared to one pair only. Sharing forces that tend to shift the system out of position, are compensated this way. In order to have a low system weight, a light beam pointer is used, which is deflected with the help of a small mirror mounted to the lower end of the shaft. An eddy current element provides damping. Double magnetic shielding protects the instrument from disturbing magnetic fields. The outer shield consists of a material which needs a very high induction for saturating it and the inner shield material provides a high permeability. All inner system components consist of non-magnetic materials except for the soft iron vanes. This ensures a very symmetrical coil field.

For linearity reasons the shaft movement $\Delta\alpha$ for full scale deflection is designed to be small. But a wide scale range is obtained by employing a

Fig. 5.3-1. **Moving Iron, Instrument as Panel Meter**
(*Hartmann and Braun, Germany*).
Simple Non-linear Construction

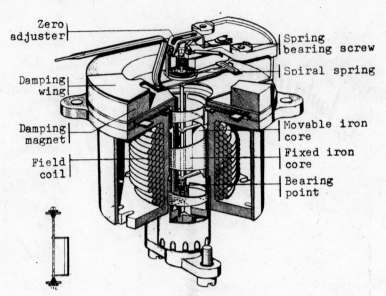

Fig. 5.3-2. Moving Iron Instrument as Panel Meter
(*Hartmann and Braun, Germany*).
Linear Wide Angle Instrument

light beam as pointer which is reflected several times this way extending its length considerably (Fig. 5.3–3). Two fixed mirrors serve this purpose.

5.4 SCALE EQUATION OF SOFT IRON INSTRUMENT

For a linear displacement energy is defined as integral of the force effecting the movement along the covered distance. So, for a circular movement, as presumed here, the mechanical energy E_m is certainly the integral of the mechanical torque $T_m = C \cdot \alpha$ (provided by the restoring spiral spring) over the angular displacement from 0 to α:

$$E_m = \int_0^\alpha T_m(\alpha)\, d\alpha$$

$$\frac{d E_m}{d\alpha} = T_m(\alpha)$$

The energy which causes the deflection α is contained in the magetic field of the coil which is (chapter 2.1.4)

$$E(\alpha) = \tfrac{1}{2} L(\alpha)\, I^2$$

This formula assumes that the inductance of the coil $L(\alpha)$ is a function of α, which is the case for instruments with concentric vanes as in Fig. 5.3-2. This energy may be introduced into $T_m(\alpha)$, for, in the state of equilibrium of the electrical torque and the mechanical one, the magnetic energy $E(\alpha)$ is

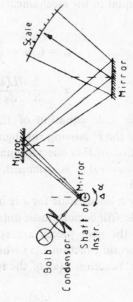

Extending the pointer length by folding the light beam.

Hartmann and Braun, Germany

Fig. 5.3–3. Precision Soft Iron Instrument.

certainly equal to the mechanical energy E_m of the spring which has absorbed this energy:

$$T_m = C \cdot \alpha = \frac{1}{2} \frac{dL(\alpha)}{d\alpha} I^2$$

$$\alpha = \frac{1}{2C} \frac{dL(\alpha)}{d\alpha} I^2$$

This is the scale equation of the moving iron system. It shows that it is sensitive to the I^2 passing through the coil. If I is an alternating current i an integration of i^2 is effected due to the torque inertia of the moving parts, and its mean value is obtained. This is the squared RMS value I^2, see chapter 3.1.

Usually a linear scale for I is needed. But the instrument responds to I^2. So the scale will be non-linear unless certain measures are taken. As shown in Fig. 5.3-2 the iron pieces are specially shaped for this purpose. They ensure a certain function $L(\alpha)$ which provides linearity of the indication. A logarithmic function $L(\alpha)$ of the inductance serves our needs (Fig. 5.4-1)

$$L(\alpha) = L_1 + L_0 \ln \frac{\alpha}{\alpha_0}$$

$$\frac{dL(\alpha)}{d\alpha} = L_0 \cdot \frac{\alpha_0}{\alpha} \cdot$$

$L_1 =$ Inductance for zero deflection α_0.

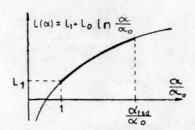

Fig. 5.4-1. Logarithmic Coil Inductance of a Moving Iron Instrument.

Introducing this expression into the scale equation produces

$$\alpha = \frac{\alpha_0}{2C} \frac{L_0}{\alpha} \cdot I^2$$

$$\alpha^2 = \alpha_0 \frac{L_0}{2C} I^2$$

$$\alpha = \sqrt{\alpha_0 \frac{L_0}{2C}} \cdot I$$

Indeed, a linear scale is obtained. Still the instrument is sensitive to the RMS value I, as the action principle has not changed.

6 Properties of Other Instruments

6.1 DIFFERENT MOVING SYSTEMS

Few more instruments, other than the moving coil and moving iron systems, are of good use occasionally. They are the moving magnet system, electrostatic instruments, bimetal instruments, vibration instruments, etc.

6.1.1 Moving Magnet System

The moving magnet system is usually a light instrument with the feature that the moving parts do not need an electrical connection. The outer coil carries the current to be measured. The action principle of the moving coil system is employed, but it is reversed: The permanent magnet is a magnetized movable disc (2) and the coil (3) is rigidly fixed. It surrounds the disc (2). The restoring torque is produced by a spring (5). The blade (1) provides the damping, see Fig. 6.1.1-1.

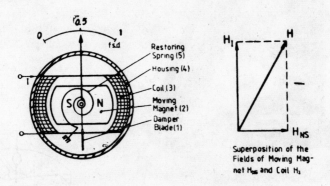

Fig. 6.1.1-1. Moving Magnet Instrument.

The zero position of the pointer is fixed by the spring (5). If the coil (3) is powered, it generates a field H_I perpendicular to the windings. H_I tends to deflect the movable disc (2) which provides its own field strength H_{NS} being superimposed to H_I. The direction of the resulting vector H gives the reading.

6.1.2   Electrostatic Measuring Instruments

These instruments are designed like capacitors, consisting of a fixed and a movable plate. According to **Coulomb's law** the latter is attracted by the force F due to the electrostatic charges Q_1 and Q_2 on the plates and their distance d: $F = K \cdot Q_1 \cdot Q_2/d^2$. A high tension instrument of this design is shown in Fig. 6.1.2-1 (a). A movable plate 2 is suspended with the help of two strips which are fixed to the upper rod. Together with two fixed plates (1) the field is applied. For this purpose the inner plate (2) is connected to one of the fixed plates (1). This way a repulsion force effects plate (2) (like charges). But opposite charges of the other fixed plate (1) towards (2) cause an attraction force which supports the other force. The displacement of plate (2) however is quite poor. So a step up mechanism (3) is employed to achieve a good pointer movement. Other instruments such as the quadrant electrometer, [Fig. 6.1.2-1 (b)], are of different design, but make use of the same principle which is selfevident from the depiction.

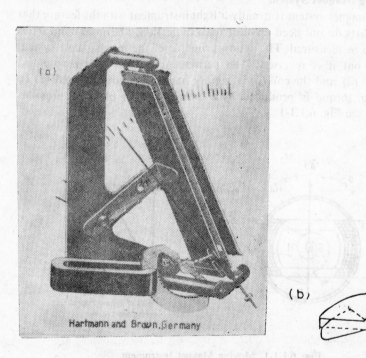

(a)

Hartmann and Brown, Germany

(b)

Fig. 6.1.2-1.   Electrostatic High Tension Voltmeters.

Sandwich designs using the latter type 18 fold, one upon the other, help to improve the sensitivity considerably. 100 V instruments of class 0.5 are obtainable. Their advantage is a very high input impedance of several $G\Omega$— Due to the construction, the repelling charge Q_1 equals the attracting one Q_2. They are both proportional to the voltage to be measured. So, the deflecting force F will be proportional to U^2. This shows that electrometers are sensitive to the squared RMS value. But U is indicated, so the scale is non-linear.

6.1.3 Bimetal Instruments

Bimetal instruments make use of a sensing strip which is composed of two metal sheets of different temperature coefficients of their expansion. They are welded to each other. Applying heat to the strip causes one metal to expand more than the other, thus bending the device.

In order to obtain a good deflection a long bimetal strip is usually wound as a spiral of several windings [Fig. 6.1.3–1 (a)].

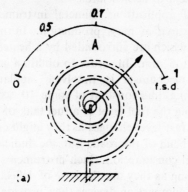

(a)

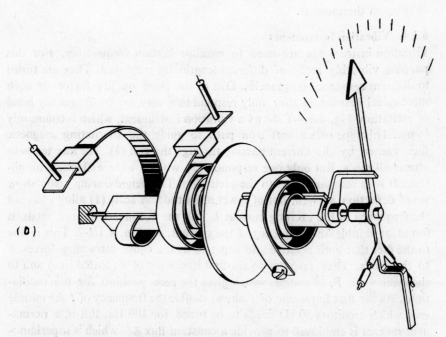

(b)

Fig. 6.1.3–1. Bimetal Instrument. (*Hartmann and Braun, Germany*)
(a) Action Principle (b) Ammeter

Fig. 6.1.3-1 (b) makes use of this principle for an ammeter. The left spiral strip senses the current to be measured, which passes the strip thus heating it up. The right spiral strip acts as a temperature compensation element. Its winding sense is opposite to the sensing spiral, so environmental temperature changes cause opposing torques thus compensating each other.—The measured current should not effect the right strip. This is ensured by the shielding plate between both spirals which protects the compensation element from the sensed heat of the left spiral. Due to considerable internal friction no separate damper is needed.

The instrument senses the current heat. That means it is sensitive to the RMS value of the current to be measured

There are plenty of applications. Bimetal instruments for automotive vehicles, are manufactured in mass production. In this case, they employ straight bimetal strips which are surrounded by a heater. The deflection of a 4 cm strip is about 5 mm. In order to obtain a good deflection of the indicator a step up mechanism is involved similar to the one of Fig. 6.1.2-1 (a). The long time constant of some 10 scconds is advantageously used for indicating the fuel level in a fuel tank of a car. The indicator cannot respond to the fast level changes as they might occur due to potholes or following a narrow bend of the road. But the indicator can follow the slow change due to fuel consumption. Such instruments are low cost devices of poor accuracy, as long as they are the base of deflection instruments. But for trigger switches or comparison devices they provide sufficient accuracy, as in room thermostats.

6.1.4 Vibration Instruments

Vibration instruments are used to monitor certain frequencies. For this purpose vibrating reeds of different lengths are employed. They are tuned to different resonance frequencies. Due to the good quality factor of such mechanical resonators they only respond to a very narrow frequency band of excitation. Fig. 6.1.4-1 shows a vibration instrument, which is commonly in use. This type uses a soft iron path to guide the alternating magnetic flux, caused by the current passing through the coil (3). The flux tends to attract all reeds. But only one responds (2) with a wide deflection amplitude. It is in resonance due to the excitation. The neighbouring reeds show some deflection as well, but of lower amplitude. A scale (1) allows to read the frequency of the exciting current I. But the resonance of the reeds is found at double the frequency of the current, see Fig. 6.1.4-2. This is due to the fact that both positive and negative fluxes cause attracting forces F to the reeds. They respond to the first harmonic of F (dotted line) and to the mean value $\bar{F}$, of course, which gives the zero position for the oscillations. As the first harmonic of F shows double the frequency of I the middle reed which monitors 50 Hz needs to be tuned for 100 Hz. But if a permanent magnet is employed to provide a constant flux $\phi-$ which is superimposed to the alternating flux ϕ, the force F is of the same frequency as I.

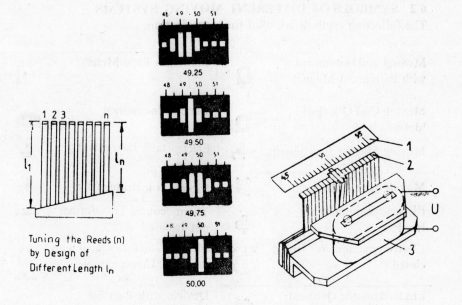

Fig. 6.1.4-1. Vibration Instrument and Typical Indications.

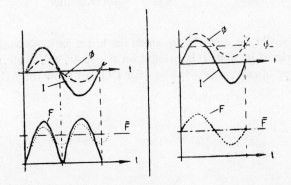

Fig. 6.1.4-2. Force *F* in Vibration Instruments without Permanent Flux and with Permanent Flux ϕ_-.

In this case the reeds need to be tuned to the frequency which they are to monitor.—The first construction is of shorter height than the second one, because the latter needs double the length of the reeds.

6.2 SYMBOLS OF DIFFERENT MOVING SYSTEMS
The following symbols are used for identification.

Moving coil Instrument with Permanent Magnet		Induction Type Meter	
Moving Coil Quotient Meter		Bimetal Instrument	
Moving Magnet Instrument		Electrostatic Instrument	
Moving Iron Instrument		Vibration Instrument	
Electrodynamometer Ironless		Thermocouple Transformer not Isolated	
Electrodynamometer Iron closed		Moving Coil Instrument with Isolated Thermocouple	
Electrodynamic Quotient Meter, Ironless		Devices with Rectifier	
Electrodynamic Quotient Meter, Iron Closed		Moving Coil Instrument with Rectifier	

Additionally the quantity to which the instrument responds is signified with the help of its unit. It is printed on the scale, (V, A etc.).

The type of current is given as—for DC and as $\sim$ for AC.

7 Compensation Instruments

Compensation instruments are also called *null deflection instruments.* They compare the unknown quantity with a known one of the same kind. For this purpose the difference of the two quantities is measured and the known quantity is changed to achieve zero for the difference. Once zero is obtained the known quantity equals the unknown one. A compensation network provides the known quantity. Commonly, it may be changed throughout wide ranges. Especially for small measurands, such as the e.m.f. of a thermocouple, the null method is superior to the deflection method, which cannot provide sufficient sensitivity. The comparison network usually consists of resistors, which may provide good accuracy for a reasonable price. They determine essentially the total accuracy of the compensation device, which is therefore, normally high. The compensated instrument does not load the measurand.

7.1 VOLTAGE COMPENSATORS

7.1.1 Action Principle

Fig. 7.1.1-1 depicts the action principle of a voltage compensator as, it was invented by POGGENDORF. In order to obtain zero voltage across the galvanometer, the slider position of the standard resistor R_S needs to be changed. The auxiliary current I_A, supplied by the standard voltage source U_S causes a voltage drop across R. It is equal to the unknown voltage U_X, for the balanced condition.

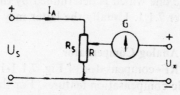

Fig. 7.1.1-1. Action Principle of a Voltage Compensator.

$$U_X = I_A \cdot R = \frac{U_S}{R_S} \cdot R$$

As I_A is constant, the part R of R_S is proportional to the voltage to be measured. It is calibrated in volts, providing the reading of U_x.

The balanced compensator does not draw current from U_X. There is no potential difference across G. So no current can flow and the input impedance of the instrument is (nearly) infinity. No loading of U_X takes place. We have an ideal voltmeter.

Ex.   The relative measurement error ϵ depends on the voltage to be measured, of course. Assume the resolution of the galvanometer G is $\Delta U = 10\mu$ V. which at least needs to be available to sense a deviation from zero. Say $U_X = 1\,V$. So ϵ_1 is obtained as $\epsilon_1 = \Delta U / U_X = 10^{-5}$. But for $U_X = 100$ V ϵ_{100} is found as $\epsilon_{100} = 10^{-7}$.

7.1.2   Technical Compensator

LINDECK and ROTHE modified the principle of Fig. 7.1.1-1 and designed the technical compensator, (Fig. 7.1.2-1). It has a fixed standard resistor R_S, but the auxiliary current I_A is changed employing R_A. As I_A is measured the voltage U needs, not be supplied by a standard voltage source. After balancing, the voltage drop across R_S equals U_X.

$$U_X = R_S \cdot I_A$$

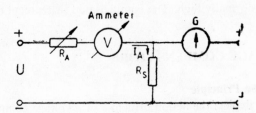

7.1.2-1.   Technical Compensator.

The scale of the ammeter is calibrated in volts to allow a direct reading of U_X. The accuracy of the technical compensator is limited to that of the ammeter, minus the one which is determined by the resolution of the galvanometer, see chapter 7.1.1. Usually the latter may be neglected.

7.1.3   Automatic Analog Compensator

The POGGENDORF–compensator of Fig. 7.1.1-1 may easily be changed to obtain automatic compensation features. For this purpose the galvanometer is replaced by a sensitive null motor M which senses the difference between the unknown voltage U_X and the voltage drop across the known part of the potentiometer P, (Fig. 7.1.3-1).

If the voltages are not equal the motor M is powered and changes the slider position until the balance point is reached. The voltage drop across M will be zero then and the motor comes to a standstill. Note that the polarity of the motor connection should not be reversed.

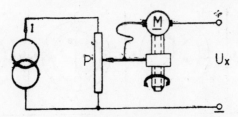

Fig. 7.1.3-1. Action Principle of an Automatic
Analog Compensator.

Compensograph

The **COMPENSOGRAPH** III of SIEMENS makes use of this principle. In
order to have sufficient sensitivity, the difference signal ΔU which is needed
to power the motor is amplified. As it is easier to amplify ACs than DCs,
ΔU is converted into an AC signal first. Consequently M is an AC-motor
(Fig. 7.1.3-2).

A bridge is employed to compose the difference $\Delta U = U_X - U_C$. As
long as ΔU is different from zero the motor M causes the potentiometer
slider to move. Its final position provides the reading of the measurand
U_X. Fig. 7.1.3-2 employs a bridge in order to have the feature of U_X being
positive or negative. This causes deviations of the slider from its middle
position to the right or to the left respectively. Usually the slider holds a
pen which allows to write a time dependent trace of U_X on a paper strip.
A feed drive provides for the motion of the recorder chart proportional to
time. The potentiometer length equals the width of the strip.

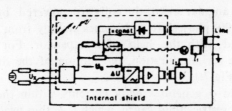

Fig. 7.1.3-2. Compensograph III, SIEMENS.

A highly appreciated feature needs to be mentioned here: Dust or other
layers on the potentiometer wire usually tend to disturb the performance
of the instrument. But as the slider feeds a constant current I into the
bridge, this current remains constant even if such disturbances are present.
So the voltage U_C remains constant and the reading is not effected. Even
thermo electric voltages are rejected. They might easily arise due to the
fast slider motions (up to 2 m/s) which may cause high temperatures due
to friction losses. But even then the current I remains constant.

The pen drive of the compensograph III is effected by a robust 2-phase
asynchronous motor. Its stator is excited by two coils which are perpendicular to each other. In order to obtain a rotating flux ϕ of constant amp-

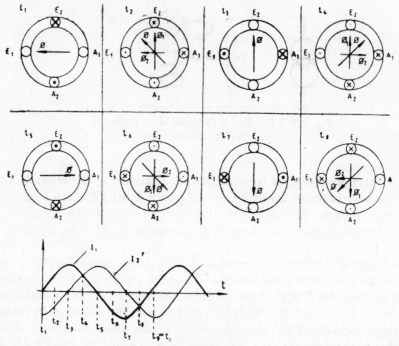

Fig. 7.1.3-3. Rotating Field Generation in a 2-Phase Motor.

Coil 1 : A_1 and E_1.

Coil 2 : A_2 and E_2.

litude and constant angular speed, the coils are powered by two currents I_1 and I_2 which are 90° phaseshifted. I_1 comes directly from the line. I_2 is usually phase shifted with the help of a capacitor. For different time instants, during one period duration, the motion of the flux vector ϕ is vividly depicted in Fig 7.1.3-3.

The rotating field ϕ generates eddy currents within the short-circuit-rotor. They tend to restrict their cause, which is ϕ, thus producing the rotation of the armature, see also chapter 4: Induction type work meter. The eddy current disc is driven likewise.

The compensograph XY of SIEMENS is a similar instrument. But it uses a DC motor, which allows very fast responses to changes of U_x (Fig. 7.1.3-4). However, the amplification of the difference signal ΔU employs AC amplifiers. This is due to the fact that a very high gain which is not effected by any drift can only be realized by employing an AC amplifier (2). Block (1) transforms the DC into an AC. In order to maintain the sign of ΔU, needed for the direction of the motor revolutions, the rectifier (3) needs to be phase sensitive. Details of a phase discriminator are given in chapter 16. The buffer (4) provides the drive power.

A constant current I feeds the bridge to reject thermovoltages, a changing contact resistance of the slider, etc. The potentiometer for U_C allows to adjust the zero position of the pen within the whole indication range. The capacitor speeds up the drive immediately once a U_x-change takes place. It cares for an exaggerated ΔU if a change of U_x occurs, this way commanding a fast response of the instrument.

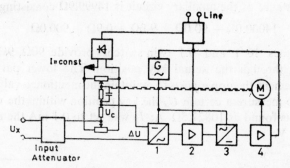

Fig. 7.1.3–4. (SIEMENS) Compensograph XY (x-channel depicted).

The fig. 7.1.3-4 depicts the x-channel, but quite the same device is provided for the y-channel making the compensograph an x-y- plotter.

7.1.4 Manual Digital Compensation
The FEUSSNER-compensator of Fig. 7.1.4–1 replaces the potentiometer by switches as shown in Fig. 7.1.4–2 increasing the accuracy of the device considerably due to precision resistors. As they may be changed in steps only, this method is of digital nature.

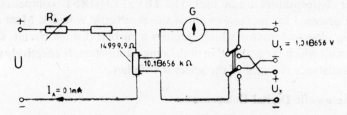

Fig. 7.1.4–1. Action Principle of the FEUSSNER-Digital-Compensator.

Before starting the measurement the auxiliary current I_A is set to exactly 0.1 mA. This can be done with the help of R_A and by employing an exact standard voltage of $U_S = 1.018656$ V as it may be provided from a WESTON standard battery. The total resistance of all resistors equals 14999.9Ω. But with the help of the switches 10.18656 KΩ are selected. R_A allows to set the current I_A to exactly 0.1 mA. This can be monitored employing the galvanometer G. It produces zero indication if U_S equals the voltage drop across the switches. This is the case only for 0.10000 mA.

For measurements the input of the instrument is switched over to U_x. Usually U_x is different from U_S. So the position of the resistor switches needs to be changed in order to produce zero indication of *G*. The key position gives directly the reading of U_x. It is the actual resistance across the selector switches times 0.1 mA.

The technical realization of this instrument is shown in Fig. 7.1.4–2. The total resistance of the auxiliary circuit is 14999.9Ω consisting of

$$14000.0Ω + 90.0Ω + 9.0Ω + 0.9Ω + 900.0Ω$$

For the decades 10, 1 and 0.1 twin switches provide 90Ω, 9Ω and 0.9Ω independently through the actual key position. The lower parts of these decades supplement the upper parts to maintain the mentioned values of their resistances. To measure a certain U_x the key position within the compensation circuit was found as 10827.5Ω. As I_A was set to 0.1 mA the measurand $U_x = 1.08275$ V.

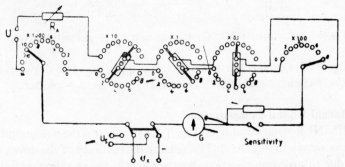

Fig. 7.1.4–2.　Voltage Compensator (FEUSSNER).

Other compensators are in use. The **DIESELHORST** compensator is specially designed for measurements of thermoelectric voltages. Most useful is the cascade-compensator which employs a current divider circuit similar to the circuit which is employed in multi-range-ammeters. It effectively rejects contact resistance changes of the selector switches.

7.1.5　Automatic Digital Compensator

The compensation could be effected automatically. The ohm-meter of Fig. 7.1.5–1 serves as an example. The compensator works stepwise and it compensates the voltage U_x across R_x. The compensation voltage is obtained as voltage drop U_S across the standard resistor R_S. In case U_S is smaller than U_x an error signal $(U_S - U_x)$ is sensed with the help of the zero amplifier. Via the voltage controlled oscillator (VCO) this difference is transferred into a pulse signal. The frequency of it is proportional to $(U_S - U_x)$. The arriving pulses are added up by the counter decades. They switch more and more resistors into action, in such a way increasing the current through R_S, which increases the voltage U_S. This procedure decreases gradually the difference $(U_S - U_x)$ until it becomes zero. In this case no analog signal reaches the VCO and so its output frequency will be zero: no pulses are generated.—For a positive error signal (+) all incoming pulses are added up,

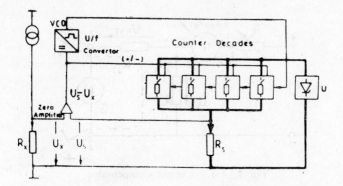

Fig. 7.1.5–1. Automatic Step Compensator as Ohm Meter.

as described. But if U_S was greater than U_x a negative $(-)$ signal for the counting direction appears. So the counter resistors are gradually switched off, thus decreasing the current through R_S and the related voltage U_S as well, until $(U_S - U_x) = 0.$—The compensation is effected once. Only a change of U_x will cause another cycle. In this way, the indication remains stable until a new measurement needs to be done.

The voltage U_x is proportional to the unknown resistance R_x as it is supplied by a constant current. So the calibration of the instrument will be given in Ohms right away.

The counter decades are integrated circuits of the state-of-the-art, also the VCO. Digital display units (not depicted) allow the presentation of the measurand, see chapter 12. For the resistors of the decades, precise components are needed. But of course the precision of major decade resistors needs to be higher than for lower decades to ensure that a change of the least significant bit, will not be covered by the tolerances of the more significant bits ($\Delta R/R \leqslant 1/2^n$ for the nth bit, see also chapter 12).

7.2 CURRENT COMPENSATOR

While a voltage compensator measures a voltage without drawing a current the current compensator allows to measure a current without a voltage drop across its terminals. The instrument proves to provide an internal impedance which is zero after balancing.

7.2.1 Action Principle

The action principle of a current compensator is shown in Fig. 7.2.1–1.

I_x is passed through the terminals 1 and 2. Certainly a deflection of the galvanometer G may take place. But the device can be balanced with the help of R_A which allows to change the auxiliary current I_A. In this way the voltage drop U_{RS1}, across the standard resistor R_{S1}, can be made equal to U_{RS2} across R_{S2}. Due to the direction of these voltages the voltage drop across G will then be zero.

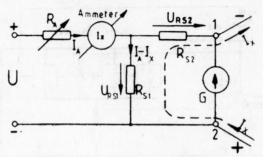

Fig. 7.2.1-1.   Current Compensator.

$$U_{RS1} = U_{RS2}$$

$$(I_A - I_x)\, R_{S1} = I_x\, R_{S2}$$

$$I_x = \frac{R_{S1}}{R_{S1} + R_{S2}}\, I_A$$

I_A is proportional to the measurand I_x. To obtain a direct reading of I_x the ammeter is calibrated in terms of I_x.

The current compensator may be an analog device or a digital step compensator. The compensation may be effected manually or automatically. For details see the previous chapter.

7.2.2 Current Compensator as Ohm Meter

The current compensator may be employed to assemble a linear ohm meter. For this purpose the unknown current I_x of Fig. 7.2.1-1 is replaced by a constant known standard current I_S and the known standard resistor R_{S2} by the unknown component R_x (Fig. 7.2.2-1).

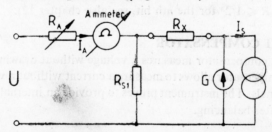

Fig. 7.2.2-1.   Current Compensator as Linear Ohm-Meter.

The last equation becomes

$$I_S = I_A \frac{R_{S1}}{R_{S1} + R_X}$$

$$R_X = \left(\frac{I_A}{I_S} - 1\right) R_{S1}$$

The measured current I_A shows a linear relationship to the measurand R_x. The scale of the ammeter may be calibrated in terms of the unknown resistance R_x, that means in ohms.

7.3 DC MEASUREMENT BRIDGES

Bridges are commonly in use to determine the resistance of the component under test. Like compensators they use galvanometers to find the balance point.

7.3.1 Wheatstone Bridge

WHEATSTONE designed the basic bridge in 1843. It became fundamental for all other bridges.

Circuit of the Bridge and Bridge Determinant

The circuit of the WHEATSTONE-bridge consists of four resistors. One of them may be unknown, for example $R_1 = R_x$. The bridge is supplied with a voltage U. It provides the branch currents I_1 and I_2. They cause voltage

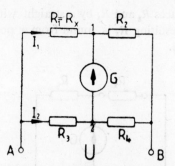

Fig. 7.3.1–1. Wheatstone Bridge.

drops across R_1 and R_3. Their difference is sensed by the galvanometer G. If the resistors R_2, R_3 and R_4 are arranged in such a way as to produce zero deflection of G the voltage drops across R_1 and R_3 are equal as ¡also the ones across R_2 and R_4:

$$I_1 R_1 = I_2 R_3$$

$$I_1 R_2 = I_2 R_4$$

Dividing these equations through each other produces

$$\frac{R_1}{R_2} = \frac{R_3}{R_4}$$

$$R_1 R_4 - R_3 R_2 = 0$$

Arranging the last equation the way as presented here shows that it is the solution of the so-called bridge determinant BD.

$$BD = \begin{vmatrix} R_1 & R_2 \\ R_3 & R_4 \end{vmatrix} = 0$$

The elements of BD are arrangad just the same way as they are positioned within the bridge. Knowing this, the unknown component $R_1 = R_x$ may easily be calculated:

$$R_x = R_2 \frac{R_3}{R_4}$$

The supply voltage U does not appear. So U may change even during the measurement. It does not effect the result. Only the accuracy of the zero reading of the galvanometer is concerned with U: If a slight deflection from zero was still remaining then this certainly may be increased by choosing a higher U. In this way a high U provides the possibility of a more accurate balancing.

7.3.2 Slide Wire Bridge

The slide wire bridge replaces R_3 and R_4 by a straight wire of constant cross section area A, and the resulting R_x is found from a position measurement of the slider (Fig. 7.3.2-1).

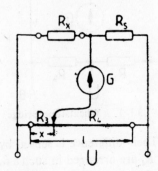

Fig. 7.3.2-1   Slide Wire Bridge.

Determination Equation

For a certain slider position x, the balance point of the bridge may be obtained. The wire length x gives the resistance R_3 and the length $(1-x)$ gives R_4:

$$R_3 = \rho \frac{x}{A} \qquad R_4 = \rho \frac{(l-x)}{A}$$

For the balance point was

$$\frac{R_1}{R_2} = \frac{R_3}{R_4} = \frac{x}{l-x}$$

This equation determines the unknown resistance $R_1 = R_x$ as

$$R_x = R_2 \frac{x}{l-x}$$

R_2 is a standard resistor R_s. It is used for selecting the measurement range. It should be chosen in such a way as to allow the reading of x near the middle position of the slider. This ensures a low measurement error.

Errors due to Uncertainties of the Slider Position and the Standard Resistor: The last equation allows to determine the relative error $\Delta R_x / R_x$ due to uncertainties Δx of the reading x. The "total difference" which provides the absolute error ΔR_x permits a fast calculation procedure. It allows to consider the tolerance $\Delta R_s / R_s$ of the standard resistor, too. It was

$$R_x = \frac{x}{l-x} R_s$$

$$\Delta R_x = \frac{\partial R_x}{\partial x} \Delta x + \frac{\partial R_x}{\partial R_s} \Delta R_s$$

$$\frac{\Delta R_x}{R_x} = \frac{R_s \frac{(l-x)+x}{(l-x)^2} \Delta X}{\frac{x}{l-x} R_s} + \frac{\frac{x}{l-x} \Delta R_s}{\frac{x}{l-x} R_s}$$

$$\frac{\Delta R_x}{R_x} = \frac{1}{l-x} \cdot \frac{\Delta x}{x} + \frac{\Delta R_s}{R_s}$$

The depiction of this function (Fig. 7.3.2-2), shows a minimum error for $x = 1/2$ which means middle position of the slider. The measurement range should avoid reading of x near the left or the right end of the wire.

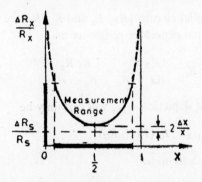

Fig. 7.3.2-2. Measurement Error over Slider Position of Slide Wire Bridge.

Error Due to Uncertainties of the Galvanometer Zero Current
Exact balance of the bridge is reached if the galvanometer current I_G is zero. But an uncertainty ΔI_G remains due to the limited sensitivity of $G \cdot \Delta I_G$ is the deviation of I_G from zero which does not allow any assessment

whether a deflection of the galvanometer pointer has taken place or not. The effect of this current error $\Delta I_G = I_G - 0 = I_G$ on the accuracy of the measurement can be shown as follows:

Using THEVIN's theorem the Wheatstone bridge may be redrawn as depicted in Fig. 7.3.2-3. U is the supply voltage and the terminals 1 and 2 are

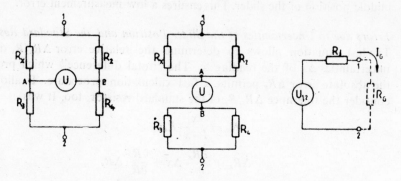

Fig. 7.3.2–3.   Equivalent Circuits of the Basic Wheatstone Bridge.

the connections for the galvanometer. The reason for giving these equivalent circuits is to show that the bridge may be considered as a generator (having the internal resistance R_i) which feeds the galvanometer (providing the load resistance of R_G). The generator e.m.f. $U_{1,2}$ may be calculated from the first depiction as

$$U_{1,2} = U \left(\frac{R_X}{R_X + R_2} - \frac{R_3}{R_3 + R_4} \right)$$

The two parallel circuits ($R_X//R_2$ and $R_3//R_4$) are in series connection and provide the internal generator resistance as

$$R_i = \frac{R_X R_2}{R_X + R_2} + \frac{R_3 R_4}{R_3 + R_4}$$

Using the last depiction $U_{1,2}$ will certainly be

$$U_{1,2} = I_G (R_i + R_G)$$

which allows to determine

$$I_G = \frac{U_{1,2}}{R_i + R_G}$$

$$= U \frac{\dfrac{R_X}{R_X + R_2} - \dfrac{R_3}{R_3 + R_4}}{\dfrac{R_X R_2}{R_X + R_2} + \dfrac{R_3 R_4}{R_3 + R_4} + R_G}$$

$$= U \frac{R_X (R_3 + R_4) - R_3 (R_X + R_2)}{\underbrace{R_X R_2 (R_3 + R_4) + R_2 R_4 (R_X + R_2) + R_G (R_X + R_2) (R_3 + R_4)}_{D}}$$

I_G is the uncertainty of the galvanometer current below which a deviation from zero cannot be detected. I_G causes an uncertainty ΔR_{XG} of R_X.

$$R_X = R_{X0} + \Delta R_{XG}$$

R_{X0} is the value of R_X as it would be found for the ideal condition that I_G really equals zero.

Now if we arrange the last equation of I_G to obtain the form of the last line for R_X we may determine the uncertainty ΔR_{XG} due to I_G. For this purpose the denominator of I_G is written as D.

$$\frac{I_G \cdot D}{U} = R_X R_4 - R_2 R_3$$

$$R_X = \underbrace{\frac{R_2 R_3}{R_4}}_{R_{X0}} + \underbrace{\frac{D}{R_4} \cdot \frac{I_G}{U}}_{\Delta R_{XG}}$$

$$= R_{X0} \left(1 + \underbrace{\frac{\Delta R_{XG}}{R_{X0}}}_{\varepsilon_{RG}} \right)$$

If a galvanometer of very high sensitivity is used ($I_G \to 0$) ΔR_{XG} approaches zero. A high supply voltage U aims for the same result. It is

$$\Delta R_{XG} = \frac{D}{R_4} \frac{I_G}{U}.$$

The total relative error of the bridge is the sum of the component errors

$$\varepsilon_{tot} = \frac{\Delta R_{Xtot}}{R_X} = \frac{\Delta R_2}{R_2} + \frac{\Delta R_3}{R_3} + \frac{\Delta R_4}{R_4} + \frac{D}{R_4} \frac{I_G}{U R_{X0}}$$

An example may be cited to facilitate the assessment of all different error causes which effect the total error ε_{tot} of the result.

Under balanced conditions the resistances of a wheatstone bridge are found as $R_2 = 9\,\Omega\,(1 \pm 0.008)$, $R_3 = 21\,\Omega\,(1 \pm 0.004)$ and $R_4 = 7\,\Omega\,(1 \pm 0.005)$. The supply voltage may be $U = 1$ V. The internal galvanometer resistance $R_G = 30\,\Omega$ and the galvanometer constant is $C_G = 10^{-6}$ A/Scd. It is possible to distinguish 0.5 scale divisions as deviation of the pointer from zero. So the remaining galvanometer current $I_G = 0.5\,\mu A$.

R_X may be determined as follows, assuming I_G being zero. So

$$R_X = R_{X0} = \frac{R_3}{R_4} R_2$$

$$= \frac{21\,\Omega}{7\,\Omega}\, 9\,\Omega = 27\,\Omega.$$

Now the relative error $\Delta R_{XG}/R_X$ may be calculated

$$\varepsilon_{RG} = \frac{\Delta R_{XG}}{R_X} = \frac{D}{R_4}\, \frac{I_G}{U \cdot R_X}$$

$$D = R_X R_2 (R_3 + R_4) + R_3 R_4 (R_X + R_2) + R_G (R_X + R_2)(R_3 + R_4)$$

$$= [27 \cdot 9\,(28) + 21 \cdot 7\,(36) + 30\,(36)\,(28)]\Omega^3$$

$$= 42336\,\Omega^3$$

$$\frac{\Delta R_{XG}}{R_X} = \frac{42336\,\Omega^3 \cdot 0.5 \cdot 10^{-6}\,\text{A}}{7\Omega \cdot 1\text{V} \cdot 27\Omega} = 1.1 \times 10^{-4}$$

The total error follows as:

$$\varepsilon_{tot} = \varepsilon_{R2} + \varepsilon_{R3} + \varepsilon_{R4} + \varepsilon_{RG}$$

$$= (0.8 + 0.4 + 0.5 + 0.011)\% = 1.711\%$$

The error term containing the galvanometer current is negligible in this case. It only needs to be considered if resistances R_2, R_3, and R_4 of higher accuracy are used.

7.3.3 Rheostat Bridge

For the previous bridges it was silently presupposed that the balance was effected by employing potentiometers which are analog devices. But the determinants may be changed in steps with the help of switches which allow to activate them discretely. The rheostat bridge of Fig. 7.3.3–1 employs such digital means.

R_2 and R_4 are designed as voltage divider. Their ratio R_2/R_4 may be 1, but it can be changed to either side to cover the ratio range from 1/1000 up to 1000/1. R_3 consists of five decades which allow R_3 to be changed between 0.1 Ohm and 10k Ohms. We have

$$\frac{R_X}{R_2} = \frac{R_3}{R_4}$$

$$R_X = \frac{R_3}{R_4} \cdot R_2$$

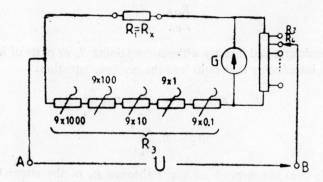

Fig. 7.3.3-1. Rheostat Bridge (Wheatstone-Type).

R_2/R_4 acts to select the range. It may be changed in decades. The numerical value of R_X is determined by R_3. The bridge covers the range of $R_X = (1 \ldots 10^6)$ Ohms. Usually that type of bridge is designed for a measurement error less than 0.02%.

7.3.4 Thompson Bridge

Resistances less than one Ohm cannot be measured, with sufficient accuracy, using the Wheatstone bridge. The resistance R_0 of the connecting wire is partly contained in the result, (Fig. 7.3.4-1).

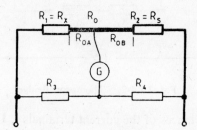

Fig. 7.3.4-1. Wheatstone Bridge: Consideration of R_0.

The upper galvanometer terminal is more or less arbitrarily connected. Assume that the connection divides R_O into the parts R_{OA} and R_{OB}. So the part R_{OA} will be considered as being contained in R_X, and the other part R_{OB} in R_S. Using the Wheatstone formula the bridge determines (instead of R_X):

$$R_X + R_{OA} = (R_S + R_{OB})\frac{R_3}{R_4}$$

In case the resistance ratio is arranged in the following way:

$$\frac{R_{OA}}{R_{OB}} = \frac{R_3}{R_4}$$

R_X can be determined directly without containing R_0 or parts of it. This is shown by introducing that ratio into the previous equation:

$$R_X + R_{OB}\frac{R_3}{R_4} = R_S\frac{R_3}{R_4} + R_{OB}\frac{R_3}{R_4}$$

$$R_X = R_S\frac{R_3}{R_4}$$

Indeed R_X does not depend on the resistance R_0 of the connecting wire between R_X and R_S. But the proper connection of the galvanometer can not practically be arranged to meet the condition of the resistance ratio. So a modification employs two additional resistors which provide the correct ratio $R_1/R_2 = R_3/R_4$. The arrangement is the Thompson-bridge (Fig. 7.3.4-2).

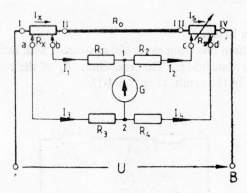

Fig. 7.3.4-2. Thompson Bridge.

The contact resistances of the current terminals I, II, III and IV would cause complications, of course. They are uncertain and variable to a high degree. But one easily gets rid of them by applying potential terminals a, b, c and d which allow a precise definition of R_X and R_S between these related terminals. In this way the contact resistances within the main current path are beyond consideration. (The contact resistances of the potential terminals are negligible compared to R_1, R_2, R_3 and R_4).

In order to determine R_X the bridge needs to be balanced. This is effected with the help of R_S. It is chosen in such a way as to make the voltage difference across the galvanometer zero. This is the case for

$$U_{1-C-IV} = U_{2-d-IV}$$

As no current can pass through G the following currents are equal:

$$I_1 = I_2 \quad \text{and} \quad I_3 = I_4$$

This means that the current through R_X equals the one through R_S: $I_X = I_S = I$. From the voltage relations, we may obtain:

$$\frac{U_{2-d-IV}}{U_{1-IV}} = \frac{R_4}{R_3 + R_4}$$

$$U_{2-d-IV} = \frac{R_4}{R_3 + R_4} \cdot I \left[R_X + R_S + \frac{(R_1 + R_2)R_O}{R_1 + R_2 + R_O} \right]$$

Similarly, we get

$$U_{1-C-IV} = I \left[R_S + \frac{R_2}{R_1 + R_2} \cdot \frac{(R_1 + R_2)R_O}{R_1 + R_2 + R_O} \right]$$

If these voltages are made equal and solution found for R_X, the result will be

$$R_X = \frac{R_3}{R_4} R_S + \frac{R_2 R_O}{R_1 + R_2 + R_O} \left(\frac{R_3}{R_4} - \frac{R_1}{R_2} \right)$$

As for the previously mentioned reasons R_1/R_2 should be equal to R_3/R_4. In this case really

$$R_X = \frac{R_3}{R_4} R_S.$$

A remaining difference of the resistance ratios in case of tolerances of R_1/R_2 and R_3/R_4 will be of minor effect if R_O is made as small as possible.

The Thompson-Bridge allows to determine resistances down to 10^{-7} Ohms. For such low values the supply current should be high, for example 100 A in order to obtain good voltage drops across R_X and R_S which actually cause the galvanometer deflection. They supply the inner bridge.

R_S is used to balance the bridge. But due to a high current passing, R_S is designed as a voltage divider. So there is no need for varying R_S. R_S remains constant. Instead the picked up voltage is changed which of course allows to balance the bridge as well.

7.3.5 Unbalanced Bridges

Bridges may advantageously be used without being balanced. In this case they are deflection bridges. The reading results from resistance changes ΔR of one resistor (Fig. 7.3.5-1), and is obtained as voltage $U_{1,2}$. The sensitivity may be doubled by using two active components, or it may even be quadrupled if all resistors are active. The last mentioned arrangement provides independence from disturbing temperatures. But all deflection bridges need a constant supply voltage U. As shown in chapter 7.3.2 the bridge behaves like its equivalent circuit of Fig. 7.3.5-1 which allows to study the loading effect of the voltmeter to the measurand U_M.

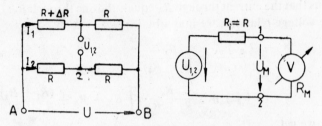

Fig. 7.3.5-1   Unbalanced Bridge and Equivalent Circuit
with Voltmeter Load

Determination Equation of an Unbalanced Quarter Bridge

$$U_{2B} = I_2 \cdot R = \frac{U}{2R} \cdot R = \frac{U}{2}$$

$$U_{1B} = I_1 \cdot R = \frac{U}{(R + \Delta R) + R} \cdot R$$

$$U_{1,2} = U_{2B} - U_{1B}$$

$$= \frac{U}{2} - U \frac{R}{2R + \Delta R}$$

$$= U \left(\frac{1}{2} - \frac{R}{2R + \Delta R} \right)$$

$$= U \left(\frac{2R + \Delta R - 2R}{4R + 2\Delta R} \right)$$

The resistance changes ΔR are usually less then 1% for instance in strain
gauge bridges (see Chapter 17.2). So $2\Delta R \ll 4R$ and it may be neglected.

$$U_{1,2} = U \frac{\Delta R}{4R}$$

This voltage is the output signal of the bridge. It depends on the supply
voltage U which should be stable.

Loading Effect of the Meter

$U_{1,2}$ is applied to the voltage divider which consists of $R_i = R$ and R_M. So
only the voltage across R_M can be measured. The voltage drop across R_i
will be lost. The load of the voltmeter decreases the sensitivity. In order
to limit the loading effect the internal resistance R_M of the voltmeter should
be high compared to R_i. If $R_M = 100 \; R_i$ only 1% sensitivity loss will be
found.

$$\frac{U_{1,2}}{U_M} = \frac{R_i + R_M}{R_M} = \frac{R + 100R}{100R} = \frac{101}{100}$$

$$\frac{U_{1,2}}{U_{1,2}} - \frac{U_{1,2}}{U_M} = 1 - \frac{101}{100} = -1\%$$

Sensitivity of Deflection Bridges

The sensitivity S is usually defined for the unloaded bridge, which produces directly $U_{1,2}$ (Fig. 7.3.5-2). S can be tested in practice by applying

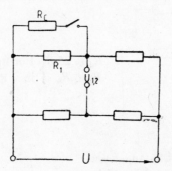

Fig. 7.3.5-2. Circuit for Determination
of the Bridge Sensitivity.

an adititional known resistor R_C parallel to R_1. This causes a resistance
change

$$\delta R = R_1 - \frac{R_1 \cdot R_C}{R_1 + R_C}$$

This value together with the obtained output voltage change $\Delta U_{1,2}$ (for
applied R_C) and the supply U allows to determine the sensitivity as

$$S = \frac{\Delta U_{1,2}}{\delta R \cdot U}$$

$$[S] = \frac{1U}{\Omega V}\left(= \frac{1}{\Omega}\right)$$

The sensitivity of deflection bridges depends on the supply voltage which
needs to be constant therefore.

Actual Bridge Circuit

Usually slight unbalances, due to component tolerances, need to be
compensated with the help of a zero potentiometer (Fig. 7.3.5-3). It allows
to balance the bridge for the case that no resistance change has taken
place. The reading across the terminals 1 and 2 will be zero. The sensitivity potentiometer allows to provide a certain part of U as supply for the
bridge. This is needed if, for instance, full scale deflection should occur for
a certain change ΔR_{max} as a consequence of the effected measurement.

The application of unbalanced bridges is appropriate wherever a resistance change is the result of quantities to be measured, for instance strain
and stress (using strain gauges, see Chapter 17.2) or temperature (using resis-

tance thermometer elements, see Chapter 17.7.2), etc. Anyway whatever quantity might be measured the unbalanced bridge performs a resistance

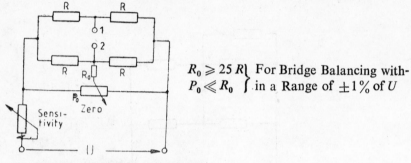

$R_0 \geqslant 25\,R$ For Bridge Balancing with-
$P_0 \ll R_0$ in a Range of $\pm 1\%$ of U

Fig. 7.3.5-3. Bridge with Adjustments
for Zero and Sensitivity.

to voltage conversion. The output voltage can directly be used for further processing. The bridge allows to measure small variations from a fixed value of the resistor under test. DC and AC applications are basically possible.

7.4 AC MEASUREMENT BRIDGES

7.4.1 Introduction
AC measurement bridges are of paramount importance in measurements. They are normally designed as rheostat bridges. Instead of a DC supply an AC supply is employed. From this fact follows that unknown impedances may be determined. They are of complex nature and are composed of an active and a reactive component. Real inductors or capacitors are of this nature. The reactive component of such an impedance depends on frequency. So an AC bridge also allows to determine an unknown frequency if the reactance is known.

7.4.2 The Complex Balancing Condition, Bridge Equation
Just as DC bridges AC bridges consist of four components (Fig. 7.4.2-1).

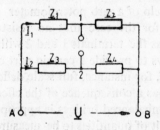

Fig. 7.4.2-1. Basic AC Bridge.

But they are complex now. The formal derivation of the bridge equation is the same as for DC bridges. In order to obtain zero voltage across the terminals 1 and 2 the impedances need to be set to values that produce equal voltages for

$$U_{A-1} = U_{A-2}$$

$$\underline{Z}_2 \cdot \underline{I}_1 = \underline{Z}_4 \cdot \underline{I}_2$$

and consequently

$$U_{1-B} = U_{2-B}$$

$$\underline{Z}_1 \cdot \underline{I} = \underline{Z}_3 \cdot \underline{I}_2$$

Dividing both equations by each other produces

$$\frac{\underline{Z}_1}{\underline{Z}_2} = \frac{\underline{Z}_3}{\underline{Z}_4}$$

or
$$\underline{Z}_1 \cdot \underline{Z}_4 = \underline{Z}_3 \cdot \underline{Z}_2 \, (*)$$

The last equation is the solution of the bridge determinant *BD* which shows the same configuration of its elements as the related circuit.

$$BD = \begin{vmatrix} \underline{Z}_1 & \underline{Z}_2 \\ \underline{Z}_3 & \underline{Z}_4 \end{vmatrix} = 0$$

$$\underline{Z} = R + jX$$

All Zs are considered as series connections of their components. But Parallel circuited components can be handled the same way if we deal with admittances.

$$BD = \begin{vmatrix} \underline{Y}_1 & \underline{Y}_2 \\ \underline{Y}_3 & \underline{Y}_4 \end{vmatrix} = 0$$

$$\underline{Y} = G + jB$$

The bridge equation (*) expresses that the products of opposite positioned impedances need to be equal in order to balance the bridge.

Need for Magnitude and Phase Balancing

Instead of writting $\overline{Z}$ as a sum of two components, the polar presentation may be used (Fig. 7.4.2-2). From the bridge equation (*) it follows:

$$Z_1 \, e^{j\varphi_1} \cdot Z_4 \, e^{j\varphi_4} = Z_3 \, e^{j\varphi_3} \cdot Z_2 \, e^{j\varphi_2}$$

$$Z_1 Z_4 \cdot e^{j(\varphi_1 + \varphi_4)} = Z_3 Z_2 \, e^{j(\varphi_3 + \varphi_2)}$$

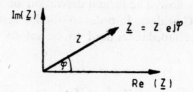

Fig. 7.4.2-2.   Polar Presentation of
an Impedance $\underline{Z}$.

From this line it may be concluded that

$$Z_1\,Z_4 = Z_3\,Z_2$$

and

$$\varphi_1 + \varphi_4 = \varphi_3 + \varphi_2$$

Referring to these equations the balancing condition may be expressed in the following way:

1. The magnitude-products of opposite positioned impedances need be equal.
2. The phase-angle-sums of opposite positioned impedances need be equal.

Or in other words: A magnitude and a phase balancing need be done. For this purpose at least two balancing keys are necessary. If there is only one zero indicator the instrument cannot distinguish whether the balance is approached from a magnitude or a phase change. In this case we may head for balance by employing a stepwise alternating procedure which is called an iterative approximation. Fig. 7.4.2-3 shows the need for magnitude and phase balancing in order to obtain the voltage $U_{1,2} = 0$ finally.

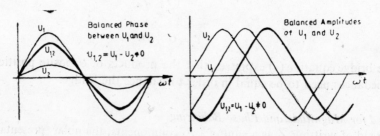

Fig. 7.4.2-3.   Diagonal Voltage $U_{1,2}$ for either Phase or Amplitude Balancing.

From the previous suggestions a practical question arises: Which impedance needs be changed with its magnitude and its phase in a known way, to determine an unknown impedance $\underline{Z}_X$ completely for its magnitude and phase ?

Compensation Using a Balancing Impedance of the Same Kind as the Unknown

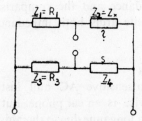

Fig. 7.4.2-4. Standard $\underline{Z}_4$ of Same King as the Unknown $\underline{Z}_x$.

Firstly, this question may be answered using Fig. 7.4.2-4. The impedances 1 and 3 are assumed as being real only. This means that $\varphi_1 = \varphi_3 = 0$. The balancing conditions become for this case as

$$R_1 \cdot \underline{Z}_4 = R_3 \cdot \underline{Z}_2 \qquad \text{and} \qquad \varphi_4 = \varphi_2$$

The phase relation shows that the balance point can only be reached by using impedances of the same kind for branch 2 and 4.

Compensation Using a Balancing Impedance of Conjugated Complex Nature as the Unknown

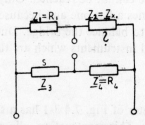

Fig. 7.4.2-5. Standard $\underline{Z}_3$ of Conjugated Complex Nature as the Unknown $\underline{Z}_X$.

Secondly the previous question can be answered employing Fig. 7.4.2-5. R_1 and R_4 are assumed to be real, so $\varphi_1 = \varphi_4 = 0$. For this case the balancing conditions are

$$R_1 R_4 = \underline{Z}_3 \underline{Z}_2$$

and
$$\varphi_3 + \varphi_2 = 0 \Leftrightarrow \varphi_3 = -\varphi_2$$

For this circuit configuration balance can only be reached if the impedances Z_2 and Z_3 are of complex conjugated nature to each other as the phase equation shows. If for instance Z_x is an inductor, Z_3 needs to be a capacitor, and vice versa.

The answer to the question above can generally be in given the following way: An unknown impedance can be determined only if in adjacent bridge arms the measured impedance and the comparison impedance are of the same kind or if in opposite arms both impedances are of complex conjugated nature.

7.4.3   AC-Null Indicators

For balancing the bridge a sensitive AC null instrument should be available. A very simple device is an ear phone. But vibration galvanometers were quite important for a long time due to their high sensitivity at resonance frequency. Nowadays they are replaced by electronic resonators. The indicator consists of a robust instrument or of a cathode ray tube.

Ear Phone as AC Null Instrument.
Ear phones are quite appropriate as null instruments. They make use of the good sensitivity of the human ear, which is frequency dependent. For 1000 Hz the sensitivity is very high. With the help of an ear phone 10^{-14} W may easily be noticed. But unfortunately the frequency of interest for AC bridges is 50 Hz. The ear sensitivity is considerably less. Only 10^{-6} W are noticeable. For industrial measuring devices this is still sufficient.—Frequently the inductance of a coil with iron core needs to be measured. Due to the nonlinear hysteresis loop, overtones are generated. But the balance of the bridge is obtainable for one frequency only which is usually the fundamental wave of 50 Hz. So a zero tone cannot be reached. Only a minimum tone can be achieved. As it is sometimes a flat one and because of disturbing noise, it becomes usually difficult to balance the bridge. Due to this fact, now-a-days one employs selective null instruments which are tuned to the fundamental frequency.

Vibration Galvanometer
The vibration galvanometer of Fig. 7.4.3-1 has a slim moving coil which is suspended by taut strips. This construction allows to set the mechanical natural frequency f_n to the frequency f_{im} of the exciting current i_m to be measured, usually 50 Hz. As the damping of such a system can be designed to be very small, the device will respond to i_m with considerable amplitude. Frequencies other than 50 Hz which might be contained in i_m, are heavily damped due to the high quality factor of such a system. The mirror on the coil allows to use a light beam for indication. For the responding galvanometer a wide line of the light trace indicates the amplitude of the coil oscillations. For the balanced bridge, the line should shrink into a spot because i_m is then zero.

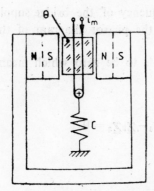

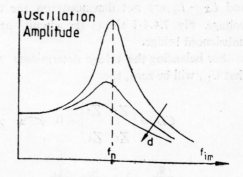

Fig. 7.4.3-1.   Mirror-Vibration-Galvanometer (Action Principle
and Frequency Response).

AC-Amplifier Galvanometers

The state-of-the-art allows design of electrically tuned circuits of the same quality factor. They are cheaper and provide good stability. These features make such instruments superior over vibration galvanometers. AC amplifier galvanometers of the low cost class may easily sense 10^{-12} W. A moving coil instrument with rectifier serves as indicator.

Null Oscilloscope

The oscilloscope allows to depict two informations at the same time. The y-channel is supplied with the bridge voltage $U_{1,2}$ and the x-channel with the bridge supply voltage U. If $U_{1,2}$ is phaseshifted towards U an ellipse will appear on the screen (see also chapter 8.4: Lissajous-Figures). In case the phase balance has already been effected an inclined straight line is depicted if the magnitude balance was not yet reached. The information about the phase is obtained from the area of the ellipse and the one about the magnitude, from the inclination angle. The handling of the bridge keys allows to meet directly both the phase and the magnitude conditions in a single attempt. A time consuming iterative balancing can be avoided in this way.

Usually null oscilloscopes are equipped with a y-amplifier of automatically controlled gain. This provides comfortable handling: For impedances which are far away from the balance point, the whole screen is used. For nearly obtained balance it is still almost fully used as well. As $U_{1,2}$ gets smaller, by approaching the balance point, the gain increases automatically. Only for deviations very close to balance, the ellipse area shrinks into a horizontal line.

7.4.4   Actual Bridges

MAXWELL's Inductance Bridge

The Maxwell bridge determines an unknown inductor for its active and reactive components of its impedance. The results of the unknowns $R_X = R_2$

and $L_X = L_2$ are not dependent on the frequency of the bridge supply voltage. Fig. 7.4.4-1 shows the circuit and the phasor diagram of the unbalanced bridge.

For balancing the bridge determinant needs to be zero which means that $U_{1,2}$ will be zero, too.

$$CD = \begin{vmatrix} \underline{Z}_1 & \underline{Z}_2 \\ \underline{Z}_3 & \underline{Z}_4 \end{vmatrix} = 0 \quad \curvearrowright \quad \underline{Z}_1\underline{Z}_4 - \underline{Z}_2\underline{Z}_3 = 0$$

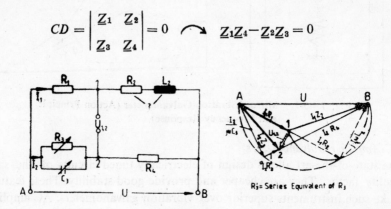

$C_3' =$ Series Equivalent of C_3

Fig. 7.4.4-1.　Inductance Bridge, Maxwell-Circuit and Phasor Diagram.

Together with the bridge components, we obtain

$$R_1 R_4 - (R_2 + j\omega L_2) \left(\frac{R_3 \cdot \dfrac{1}{j\omega C_3}}{R_3 + \dfrac{1}{j\omega C_3}} \right) = 0$$

$$R_1 R_4 + (R_2 + j\omega L_2) \left(\frac{R_3 \cdot \dfrac{j}{\omega C_3}}{R_3 - \dfrac{j}{\omega C_3}} \right) \frac{R_3 + \dfrac{j}{\omega C_3}}{R_3 + \dfrac{j}{\omega C_3}} = 0$$

$$R_1 R_4 + (R_2 + j\omega L_2) \frac{R_3 \dfrac{j}{\omega C_3} \left(R_3 + \dfrac{j}{\omega C_3} \right)}{R_3^2 + \dfrac{1}{\omega^2 C_3^2}} = 0$$

$$-\frac{R_1 R_4 \left(R_3^2 + \dfrac{1}{\omega^2 C_3^2} \right) + R_3 \left(R_2 + j\omega L_2 \right) \left(j\dfrac{R_3}{\omega^2 C_3} - \dfrac{1}{\omega^2 c_3^2} \right)}{R_3^2 + \dfrac{1}{\omega^2 C_3^2}} = 0$$

$$R_1 R_4 \left(R_3^2 + \frac{1}{\omega^2 C_3^2} \right) + R_3 \left(-\frac{R_2}{\omega^2 C_3^2} - \frac{L_2 R_3}{C_3} \right) + j R_3 \left(\frac{R_2 R_3}{\omega C_3} - \frac{L_2}{\omega C_3^2} \right) = 0$$

Real term:

$$R_3{}^2 \left(R_1\,R_4 - \frac{L_2}{c_3} \right) + \frac{1}{\omega^2 C_3{}^2} \left(R_1\,R_4 - R_2\,R_3 \right) = 0$$

$$\frac{R_3}{\omega C_3} \left(R_2\,R_3 - \frac{L_2}{c_3} \right) = 0$$

Imaginary term:

$$R_1\,R_4 = R_2\,R_3$$

and

$$R_2\,R_3 = \frac{L_2}{C_3} = R_1\,R_4$$

The first equation solves for R_2 and the second for L_2:

$$R_2 = \frac{R_1\,R_4}{R_3}$$

and

$$L_2 = R_1\,R_4\,C_3$$

R_1 and R_4 may be used as decimal range selectors (of course as $R_1 \cdot R_4$ is changed also $R_2 \cdot R_3$ needs be changed in the same way). The components R_3 and C_3 serve to determine the numerical values within the selected decades for R_2 and L_2 respectively.

The frequency is not contained in R_2 and L_2. So the bridge is not dependent on it. This is a distinct advantage for inductance measurements of coils with an iron core. Due to the non-linear hysteresis loop of iron-the diagonal voltage $U_{1,2}$ will contain higher harmonics of the supply frequency But as the bridge is balanced for all frequencies zero voltage $U_{1,2}$ can be finally reached.

Due to these frequency considerations a practical question arises: Under what conditions is a bridge not frequency dependent? This is the case if the unknown component is balanced by an impedance of opposite kind in the opposite branch. This is realized within the Maxwell bridge. $\underline{Z}_3$ needed to be of opposite circuit, which was a parallel circuit of R_3 and C_3 as $\underline{Z}_2$ was a series connection of R_2 and L_2. But independence from frequency occurs as well if the balance is effected with the help of a standard of the same kind in adjacent arms. The kind of circuit neeeds to be the same as the one of the device under test. That means: The series inductance in branch 2 can also be determined with the help of a standard series inductance in branch 4.

Ex. If $R_1 = R_4 = 1$ K Ohm are chosen ($R_1 \cdot R_4 = 10^6\ \Omega^2$) the change of C_3 should be effected in μF. In this case $L_x = L_2$ can be directly read in Henry from the key of C_3:

$$[L_2] = [R_1 \cdot R_4] \, [C_3]$$

$$1\text{H} = 10^6 \frac{V^2}{A^2} \cdot 10^{-6} \frac{As}{V} = 1 \frac{Vs}{A}$$

$R_2 = R_x$ may be found from an inversed scale of R_3 in Ohms, or if the bridge is designed as a rheostat bridge the resistors of R_3 may be inversely connected to the selector key of R_3. In case R_3 is changed in MOhms, R_2 may directly be read from the key of R_3 in Ohms:

$$[R_2] = \frac{[R_1 \cdot R_4]}{[R_3]}$$

$$1\Omega = \frac{10^6 \Omega^2}{10^6 \Omega} .$$

Capacitance Bridge of WIEN

The Wien bridge is used for determining the capacitance and the loss resistance of an unknown capacitor. The circuit is based on the series equivalent for the unknown component. In this configuration a fixed standard (S) may be used as comparison component C_1. The magnitude balance is effected with the help of R_1 and the phase balance by R_3 (Fig. 7.4.4-2).

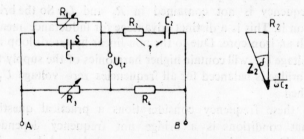

Fig. 7.4.4-2.   Capacitance Bridge (Wien).

Following the same procedure as for the Maxwell bridge, the calculations result in

$$C_2 = C_1 \frac{R_3}{R_4} \cdot \frac{(\omega C_1 \, R_1)^2 + 1}{(\omega C_1 \, R_1)^2} = f(\omega)$$

$$R_2 = \frac{R_4}{R_3} \frac{R_1}{(\omega C_1 R_1)^2 + 1} = f(\omega)$$

$$\tan \delta_2 = R_2 \omega C_2 = \frac{1}{\omega C_1 R_1}$$

For the realistic case that

$$R_1 \gg 1/\omega C_1 \quad (\omega C_1 R_1 \gg 1)$$

C_2 becomes quite independent of the frequency:

$$C_2 = C_1 \frac{R_3}{R_4} \neq f(\omega)$$

But the loss resistance R_2 (as well as the tan δ_2) depends on ω. So the bridge should have a stable frequency supply.

Frequency Bridge of ROBINSON

Robinson used the frequency dependance of the Wien bridge for frequency measurements. For this purpose he chose the bridge component as follows:

$$R_4 = 2R_3$$

$$C_2 = C_1 \frac{R_3}{R_4} = C_1 \frac{R_3}{2R_3} = \frac{C_1}{2}$$

$$R_1 = R_2 = R$$

$$R = \frac{2R_3}{R_3} \cdot \frac{R}{(\omega C_1 R)^2 + 1}$$

$$\omega^2 (C_1 R)^2 = 2 - 1$$

$$\omega = \frac{1}{C_1 R}$$

The unknown frequency $f = 1/2\pi C_1 R$ is found by changing C_1 aud $C_2 = C_1/2$ simultaneously as well as R_1 and R_2 (both equal to R) employing twin switches for the capacitors and the resistors.

Siemens designed such a bridge for the audio frequency range with an accuracy of 1%. But electronic frequency counters have replaced this device, see chapter 13.1.2.

Combined Maxwell Wien Bridge (INKAVI)

By switching over branch 1 into branch 3 a Maxwell bridge changes into a Wien bridge. The combined Inkavi bridge makes use of this fact (Fig. 7.4.4-3). It covers the ranges for L from 1μH to 10H and for C from 10pF to 100μF. The mean measurement error is around $\pm 3\%$ of the employed range.

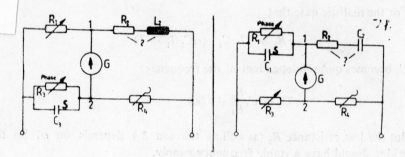

Fig. 7.4.4-3. INKAVI—Combination-Bridge (H+B, Germany)

$$L_2 = R_1 R_4 C_3 \qquad\qquad C_2 = C_1 \frac{R_3}{R_4}$$

$$R_2 = \frac{R_1 R_4}{R_3}$$

$$\tan \delta_L = \frac{R_2}{\omega L_2} \qquad\qquad R_2 = \frac{R_4}{R_3} \cdot \frac{R_1}{(\omega C_1 R_1)^2 + 1}$$

$$\tan \delta_c = R_2 \omega C_2$$

$$= \frac{R_1 R_4}{\omega R_1 R_4 C_3 R_3} \qquad\qquad = \frac{R_4}{R_3} \frac{R_1 \omega}{(\omega C_1 R_1)^2 + 1} \; C_1 \cdot \frac{R_3}{R_4}$$

$$= \frac{1}{\omega C_3 R_3} \qquad\qquad = \frac{\omega C_1 R_1}{(\omega C_1 R_1)^2 + 1} \approx \frac{1}{\omega C_1 R_1}$$

$C_3(C_1)$ is the standard capacitor. $R_3(R_1)$ is a resistance of high value. Its key is calibrated in terms of tan δ. $R_1(R_3)$ is a slide wire potentiometer which determines the value of $L_2(C_2)$. R_4 acts as decimal multiplier.

Side Note: R_1 and R_4 determine L_2 as well as R_2. Therefore, it is not feasible to determine both components in this way. This is avoided by stepping aside to tan δ instead of R_2. The tan δ does not contain R_1 and R_4.

Capacitance and tan δ Bridge (SCHERING)
The Schering bridge determines the capacitance and the loss factor tan δ of a component within the full technical range of interest (Fig. 7.4.4-4). Measurements of the dielectric constant and the ionization voltage are also frequently undertaken. This bridge is mainly used for measurements of parameters of high voltage cables. The components for operating the bridge can be designed in such a way that no high voltage can cause a danger for the operator. The bridge is grounded at the side A of the handling keys to safeguard the user. Only B is connected to the high voltage. Overvoltage indicators give a warning in case R_1 and/or R_3 are disconnected. At the

asme time these indicators provide safety, because they do not allow the voltage to exceed 120 V w.r.t. ground.

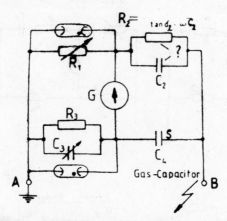

Fig. 7.4.4-4.   Schering High Tension Bridge for Cable Measurements.

The bridge calculations produce the following results

$$\underline{Z}_1 \underline{Z}_4 = \underline{Z}_2 \underline{Z}_3$$

$$R_1 \cdot \frac{1}{j\omega C_4} = \frac{\dfrac{1}{\tan\delta_2 \cdot \omega C_2} \cdot \dfrac{1}{j\omega C_2}}{\dfrac{1}{\tan\delta_2 \cdot \omega C_2} + \dfrac{1}{j\omega C_2}} \cdot \frac{R_3 \cdot \dfrac{1}{j\omega C_3}}{R_3 + \dfrac{1}{j\omega C_3}}$$

$$\tan\delta_2 = \omega C_3 R_3$$

$$C_2 = C_4 \frac{R_3}{R_1 \left[(R_3 \omega C_3)^2 + 1 \right]}$$

But for $R_3 \omega C_3 \ll 1$ becomes

$$C_2 = C_4 \frac{R_3}{R_1}$$

R_3 should be designed in such a way that it may act as decimal multiplier for tan δ. C_3 gives the numerical value.

A purchasable Schering bridge from H + B Germany, covers a C-range from 1pF to 10μF and a tan δ-range of 2.10^{-5} to 1.111 employing a standard (S) gas capacitor of $C_4 = 100$ pF. The gas provides a high pressure thus ensuring a very high break through voltage of the standard component S.

7.4.5 Universal RLC-Bridge (HP-Model 4265 A)

The Hewlett and Packard Model 4265 is a universal bridge of the state-of-the-art. It covers most applications of impedance measurements (Fig. 7.4.5-1). The circuits are occasionally modified compared to the ones we have dealt with so fas. They are recommended for further studies to the reader.

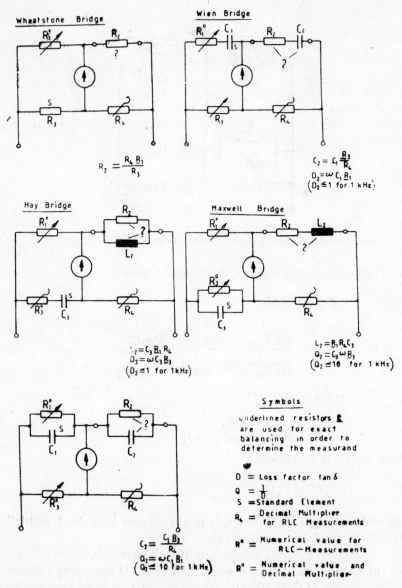

Fig. 7.4.5-1.   Choices of the HP Universal RLC-Bridge (Model 4265 A).

7.4.6   Transformer Ratio Bridge (Wayne and Kerr)

The transformer ratio bridge of Fig. 7.4.6-1, allows to determine all kinds of impedances for reactive component and parallel loss resistance. The action principle is simple. If the amperewindings within the upper and the lower coils of transformer $T2$ are equal the galvanometer G produces zero deflection because $I_x \cdot N$ compensates $I_s \cdot N$. The bridge is balanced for $\underline{I}_x = \underline{I}_s$.

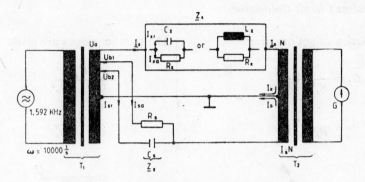

Fig. 7.4.6-1   Transformer Ratio Bridge.

As $\underline{Z}_x$ is nsually complex, I_x is composed of an active (I_{xa}) and a reactive component (I_{xr}). Both need to be separately balanced. This is effected by selecting distinct voltages U_{b1} and U_{b2} which produce the active comparison current I_{sa} and the reactive one I_{sr} of the standard combination of R_s and C_s. say $\underline{Z}_x$ is a capacitor.

$$\underline{I}_x = \underline{I}_s$$

$$I_{xa} + jI_{xr} = I_{sa} + jI_{sr}$$

$$\frac{U_a}{R_x} + \frac{U_a}{\dfrac{1}{j\omega C_x}} = \frac{U_{b1}}{R_s} + \frac{U_{b2}}{\dfrac{1}{j\omega C_s}}$$

The comparsion of the active and the reactive components produces

$$R_x = \frac{U_a}{U_{b1}} R_s \text{ and } C_x = \frac{U_{b2}}{U_a} C_s$$

For transformer $T1$, the voltage ratios are certainly the same as the winding ratios. Consequently R_x and C_x are obtained as

$$R_x = \frac{N_a}{N_{b1}} R_s \quad \text{and} \quad C_x = \frac{N_{b2}}{N_a} C_s$$

The winding ratio gives the reading.

Unknown inductors are measurable as well. Only the slider for U_{b2} needs to be taken to the lower windings of $T1$. This reverses U_{b2} which is necessary as for an inductor the current through L_x is certainly reversed, in comparison to the one through the previous capacitor C_x.

The bridge produces its readings as parallel components of the unknown impedance independently of its actual circuit. In case the series equivalent values are needed, they may be calculated from the measurement results.

Equivalent Circuit Calculations

The series values are given | The parallel values are given

$$G_p = \frac{R_s}{R_s^2 + X_s^2}$$

$$B_p = \frac{X_s}{R_s^2 + X_s^2}$$

$$R_s = \frac{G_p}{G_p^2 + B_p^2}$$

$$X_s = \frac{B_p}{G_p^2 + B_p^2}$$

$$R_p = \frac{1}{G_p} = R_s \left(1 + \frac{X_s^2}{R_s^2}\right)$$

$$= R_s \left(1 + \frac{1}{\tan^2\delta}\right)$$

$$X_p = \frac{1}{B_p} = X_s \left(1 + \tan^2\delta\right)$$

$$R_s = R_p \frac{1}{1 + \dfrac{1}{\tan^2\delta}}$$

$$X_s = X_p \frac{1}{1 + \dfrac{1}{\tan^2\delta}}$$

Dissipation factor $\tan\delta = D = \dfrac{R_s}{X_s} = \dfrac{X_p}{R_p} = \dfrac{1}{Q}$

Quality factor, Q

In order to avoid the last calculation of transforming parallel into series components and vice versa, the previous *HP*-bridge, (Fig. 7.4.5–1) allows to switch over to another mode. But this feasibility is also useful to cover wider impedance ranges.

7.4.7 Auxiliary Bridge Arms (Wagner Branch)

So far all impedances of a bridge have been considered as lumped components being found at certain local positions. However other impedances are distributed over the whole assembly. They may have considerable effect on the balance of the bridge. It might be found, for instance, that a bridge which is expected to be nullable perfectly (according to theory) can only be balanced down to a certain minimum voltage which is different from zero. This experimental result may derive from impedances, say C_A, C_B, and C_1, C_2 towards ground which may cause considerable unbalance (Fig. 7.4.7-1).

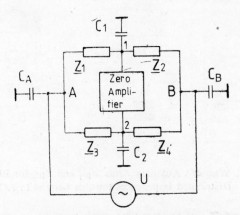

Fig. 7.4.7-1 Consideration of Capacitances C_A, C_B, C_1, C_2 of the Bridge Arms as Disturbances for Zero Balance.

The following marginal notes are designed to ease the conclusions of this chapter.—Grounding the supply U at point A shorts C_A. C_B is parallel to U in this case. This leaves C_A and C_B without effect on the bridge. But C_2 appears parallel circuited to $\underline{Z}_3$ and C_1 to $\underline{Z}_1$. Therefore, they cause measuring errors.—But grounding the zero amplifier (instead) at point 1 would short C_1. After balancing point 2 would be on ground potential, too. This measure would eliminate any effect of C_1 and C_2 on the bridge. However, C_A is parallel to $\underline{Z}_1$ then, and C_B to $\underline{Z}_2$. They would consequently produce measuring errors. These observations allow to conclude that always two disturbing capacitances are active, whatever grounding may be applicable. WAGNER'S auxiliary arms, however, provide a satisfying solution to this problem (Fig. 7.4.7–2(a)].

The auxiliary arms introduce two impedance $\underline{Z}_{W1}$ und $\underline{Z}_{W2}$. They are adjusted in such a way as to produce zero potential at their connection point W. W may or may not be grounded, nothing would change. As the capacitances of points A and B are those towards ground, C_A and C_B may be understood as being connected to the artificial ground point W. Now, adjusting $\underline{Z}_1$ the same way to obtain zero potential for point 1 allows to conclude that it does not matter whether C_A and C_B are connected to point 1 or to point W. In any case they have been cared for already by adjusting the Wagner-arms.—The adjustment of the auxiliary branch needs not be done by the user of the instrument. The manufacturer takes care of this.

One degree of freedom remains still which is made use of, to balance the bridge for determining the unknown component $\underline{Z}_2$. This is effected with

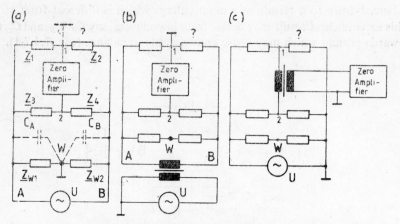

Fig. 7.4.7–2. Wagner's Auxiliary Arms Z_{W_1} and Z_{W_2} for Elimination of Distributed Impedances towards Ground in AC Bridges.

the help of $\underline{Z}_3$ and $\underline{Z}_4$, of course, also aiming at zero potential for point 2. This is mointored with the zero amplifier. The final settings of $\underline{Z}_1$, $\underline{Z}_3$ and $\underline{Z}_4$ allow to calculate $\underline{Z}_2$ for its real and its imaginary term this way providing full information of the unknown component $\underline{Z}_2$.

Now a days, AC bridges are normally supplied from electronic sources and they are monitored with electronic zero amplifiers. Both devices provide the advantage of simplicity, if they may be grounded at one end each. But the bridge should be grounded, too. All these demands are complied with in figures (b) and (c). In both circuits the unknown component is grounded, too. Circuit (c) may be the configuration of choice, because its transformer can be small as it needs to transmit only the balance information wheras the transformer of circuit (b) needs to provide the total power for supplying the complete system. It should be large, therefore, a fact which makes it costly.—Combining both circuits (b) and (c) into one by using two transformers would allow to put any point of the actual bridge to a potential of free choice, a feature which would be appreciated for measurements in high voltage applications. In this case the transformers need to be proof to high voltages to make the separation from the line safe.

7.4.8 Accuracy of Bridges
All bridges may be designed as precision instruments. They are in need of precise standards which are selected by using good switches. The zero indicator needs to provide a high sensitivity. These items are self-evident.

Stray effects between the bridge components may considerably effect the measurement. To avoid this certain precautions of shielding need to be taken. Auxiliary arms compensate distributed impedances towards ground (Chapter 7.4.7). To comply with all these needs makes precision bridges costly.

However, nowadays technology allows the design of digital automatic bridges. A step compensator of H + B, Germany, allows for instance, a class of 0.05 at a reasonable price. Such instruments are of help to classify mass produced components automatically.

8   Cathode Ray Oscilloscope

8.1   GENERAL FEATURES

The cathode ray oscilloscope makes use of an electron beam, inside an evacuated glass bulb to sense the measurand and indicate it on a screen.

The oscilloscope is an electrostatic measurement device. The movable pointer is composed of electrons being subjected to the applied electrostatic field to be measured. As electrons are the lightest charged mass particles, they can directly respond to very fast changes of the electrical field strength. 100 MHz traces may instantaneously be depicted, which is certainly an outstanding feature.

8.2   COMPONENTS AND ACTION PRINCIPLES

An oscilloscope consists of many components and functional elements, as shown in Fig. 8.2-1. The central part is the cathode ray tube. It is in need of a voltage divider which supplies the different potentials for the electron gun and for the post-acceleration system.

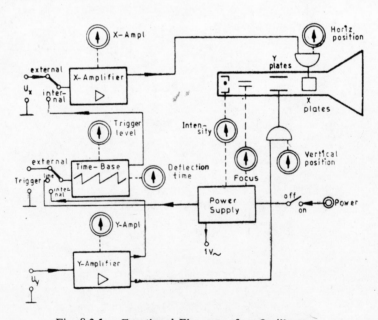

Fig. 8.2-1.   Functional Elements of an Oscilloscope.

The input attenuator is usually a part of the *x*-and the *y*-amplifier. It adopts the input voltage in such a way, as to obtain a reasonable beam deflection on the screen. The time base relates the time to the *X*-axis. The trigger feasibility provides for a standing depiction on the screen. The power supply provides all DC voltages required for all functional elements. The double circles in Fig. 8.2-1 are handling keys to adjust the scope to the measurand. They cope with wide ranges of the measured voltage for its magnitude and frequency.

8.2.1 The Cathode Ray Tube

All parts of the cathode ray tube are housed inside an evacuated glass enclosure (Fig. 8.2.1-1). It was invented in the late twenties by BRAUN.

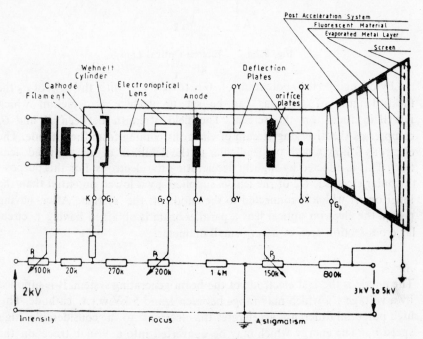

Fig. 8.2.1-1. Cathode Ray Tube.

Cathode

The cathode is indirectly heated up. Due to a barium oxide layer on the cathode surface a temperature of only 800°C is needed to emit a sufficient number of electrons per unit of time.

Wehnelt-Cylinder

The WEHNELT-cylinder is designed as an orifice plate. It is connected to a negative voltage towards the cathode which controls the beam intensity with the help of P1. In this way the brightness of the screen trace can be adjusted.

Electron Optical Lens

The electron-optical lens effects the compression of the electron beam into a thin line. The effect of the electron optical system (Fig. 8.2.1-2), is the same as the effect of a convex glass lens on visible light.

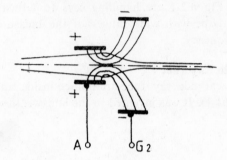

Fig. 8.2.1-2. Elctron-Optical Lens.

By changing the potential of *G2* towards *A* with the help of P_2 the focal length of the lens can be set to produce a depiction which is well in focus on the screen. The field strength between *A* and G_2 compresses the diverging stream of electrons coming from the cathode. The electrons follow a path perpendicular to the fieldlines of the applied lens field. This is effected by slightly decelerating the electrons. For this purpose the outer electrode *G2* of the lens is supplied by a lower potential than the inner one *A* which is connected to the anode of the system. After having passed the electron optical lens a parallel beam is obtained having a circular cross-section area of a very small diameter.

Anode

The anode is the last electrode of the beam generating system. It is supplied by a voltage U_A which may range between 1 and 5 kV w.r.t. cathode. This high potential difference accelerates the electrons to a considerable high speed V_A. The energy which may be converted into a visual trace on the screen is equal to the kinetic energy W_{kin} of the electrons $(m \cdot V_A^2/2)$ due to their speed which is also equal to the electrical energy $U_A \cdot e$ due to the voltage U_A which produces this energy.

Coulomb's axiom postulates a force *F* for a charge *Q* which is subject to an electric field strength E_Z $(F = Q \cdot E_Z)$. This is certainly true for an electron, too, which provides the charge of $e = 1.602 \times 10^{-19}$ As. If E_Z should change with the special location between cathode and anode, *F* would change as well. They are functions of *s* then, $E_Z(s)$ and $F(s)$:

$$F(s) = e \cdot E_z(s)$$

Mechanical work *W* is the integral of the force $F(s)$ with respect to the effected displacement *s*. Introducing $F(s)$ this integral may be considered the electrical work W_{el} which effects the kinetic energy W_{kin} of the electrons during their passage (s) from cathode to anode.

$$W = \int_0^s F(s) \, ds = e \int_0^s E_z(s) \, ds = e \, U_A = W_{el}$$

$$W_{kin} = \frac{m}{2} v_A^2 = U_A \cdot e$$

$$v_A = \sqrt{\frac{2e}{m} \cdot U_A}$$

The mass of an electron is $m = 9.1085 \times 10^{-31}$ kg. With the constants of the electron charge e and its mass m the last formula calculates the speed of the electrons once they have reached the anode.

$$v_A \left[\frac{m}{s} \right] = 594 \times 10^3 \sqrt{U_A [V]}$$

The centre of the anode is designed as a hole. The electrons are channelled through it and continue to travel with constant velocity v_A.

Deflection Plates
After having passed the anode opening, the electrons approach the uniform fields of the deflection plates. First they reach the y-plates (Fig. 8.2.1-3), which deflect the beam in vertical direction. After that they pass the x-plates, designed quite the same way but turned by 90° towards the y-plates. In this way they deflect the beam in horizontal direction. The y-plates may serve to describe the effect of the applied voltage U (to be measured) through the deflection S (to be indicated). It should be

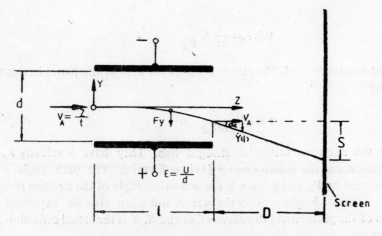

Fig. 8.2.1-3. Beam Deflection between the y-plates.

noted that the electrons which are subject to the field strength $E = U/d$ follow the same trajectory parabola as a horizontally thrown stone would do when subject to gravity.

There is no force in Z-direction effecting the electrons. Therefore, no change of speed takes place in this direction. The electrons travel with constant speed v_A through the plates. But in y-direction the field strength E effects a constant force F_y to the electrons, deflecting them towards the positive $(+)$ plate as long as they travel within the range 1 of the plates. F_y effects each electron. F_y is equal to the product of the electron charge e and the field strength E. According to Newton's fundamental law F_y is balanced by the force $m \cdot \ddot{y}$ due to the inertial mass m of the electron. The force equation determines the speed in y-direction, especially $\dot{y}$ (l) at the end (1) of the plates :

$$\underbrace{E \cdot e}_{F_y} - m \cdot \ddot{y} = 0$$

$$\ddot{y} = \frac{e}{m} E$$

$$\dot{y}(t) = \frac{e}{m} E \cdot t + C_1$$

But C_1 equals in 0 due to the initial condition that at the time instance $t = 0$ (at which the electrons enter the plate), the velocity in y-direction equals in 0_1 too.

The time dependent velocity $\dot{y}(t)$ may be transformed into a function of the location z with the help of the relation

$$V_A = \frac{z}{t} \curvearrowright t = \frac{z}{V_A}$$

$$\dot{y}(z) = \frac{e}{m} E \frac{z}{V_A}$$

At the location $z = 1$, the electron leaves the field of the plates having the velocity

$$\dot{y}(l) = \frac{el}{mV_A} \cdot E$$

Now the electrons follow a straight line. They have a velocity V_A in z-direction and the velocity and $\dot{y}$ (1) in y-direction. The path angle α is determined by V_A and $\dot{y}$ (l). α is the deflection angle of the electron pointer which excites a bright spot on the screen. But α can also be expressed in terms of the geometrical distances of D and S. S is the actual deflection on the screen.

$$\tan \alpha = \frac{\dot{y}(l)}{V_A} = \frac{el}{mV_A^2} \quad E = \frac{S}{D}$$

$$S = \frac{D\,el}{mV_A^2} \text{ with } E = \frac{U}{d}$$

$$S = \frac{D\,el}{2d \underbrace{\frac{m}{2} v_A^2}_{U_A \cdot e}} U$$

This is the scale equation of the oscilloscope. It shows that the deflection S is proportional to the voltage U to be measured. The kinetic energy $(m \cdot V_A^2/2)$ should not be too high because it decreases the sensitivity. The next equation concludes the same result: The anode voltage U_A, which accelerates the electrons to a certain velocity V_A, should not be too high.

$$S = \frac{Dl}{2d} \cdot \frac{U}{U_A}$$

The x-deflection is caused by the x-plates employing exactly the same means which result in the same deflection. So generally for both channels, we have

$$S_{x,y} = \frac{Dl}{2d} \cdot \frac{U_{x,y}}{U_A}$$

The deflection coefficient A is defined as

$$A = \frac{S_{x,y}}{U_{x,y}} = \frac{Dl}{2d} \cdot \frac{1}{U_A} [A] = 1 \frac{cm}{V}$$

Usually $1/A$ is understood as the sensitivity of the scope. It is labelled at the input attenuator keys in V/cm instead of A.

Post Acceleration System

A high velocity of the electrons would certainly a cause bright appearance of the depicted spot on the screen. A high speed is obtained from a high anode voltage U_A, which unfortunately would spoil a good deflection S. But both demands, one for good deflection and the other one for a bright depiction need be coped up with. First of all S is considered by taking U_A low. In order to get a brilliant screen appearance the beam intensity (which is the number of electrons passing a certain location within a certain time) should be increased by increasing the potential of the Wehnelt cylinder But the effect of this measure is quite limited because it tends to shift the depiction out of focus. Once too many electrons travel too near to each other, they repel each other due to the equal charges they carry. Another measure is to choose a sensitive fluorescent material for the screen layer. But best of all, is to introduce a post acceleration system (Fig. 8.2.1-4). It does not effect the deflection angle α, but only provides additional acceleration to the electrons. For this purpose concentric rings made from graphite or a spiral are introduced into the conical part of the

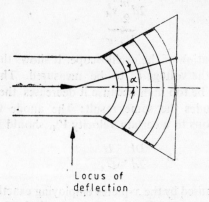

Locus of
deflection

Fig. 8.2.1-4. Post Acceleration System.

glass bulb. Their potential rises up to 5 kV. The rings provide areas of equal potentials which are ball sections, having their centre at the locus of deflection. This design allows a post acceleration independently of the deflection angle. For any α, the electrons have to pass the same difference of potentials resulting in the same final velocity for each of them. Eventually they hit the fluorescent screen with a very high speed thus producing a bright trace.

Fluorescent Screen
The screen is covered from inside with a fluorescent substance which consists of zinc compounds. They differ from each other by different persistance and different colours. Once the elelctrons hit the fluorescent material they excite it for visible light radiation in the following way (Fig. 8.2.1-5).

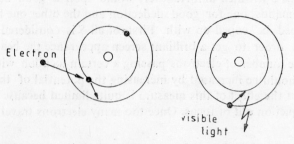

Electron

visible
light

Fig. 8.2.1-5. Excitation Mechanism of the Fluorescent Material.

The high velocity electrons hit the atoms of the light emitting substance and they lift their electrons to orbits (around the atom core) which are of a higher energy level. They stay there for a certain short time which is typical for each substance. After that they fall back to their original orbit of lower energy level. The energy difference is emitted as a light quantum $h \cdot v$ (h being PLANCK's natural constant and v, the frequency of the radiated light).

In order to avoid a negative charging of the screen trace (due to the charges of the arriving electrons) a thin metal layer of the thickness of only a few molecules is deposited on the screen from inside. It is transparent of course, but grounded in order to draw out the incoming charge.

Storage Tube
Light emitting fluorescent substances are available with wide ranges of their persistance. But fluorescent materials with a very long persistance cannot cope with non-repetitive measurement voltages, which are available only once. The screen depiction may emerge too fast for proper assessment. In such cases storage tubes are of good advantage. The principle of action is shown in Fig. 8.4.1-6. Behind the screen there is additionally a storage layer, which is positively charged. Once the incoming electrons reach this layer, they neutralize it at the location of the trace to be depicted (due to their own negative charge). According to the function $y(t)$ which the electron ray

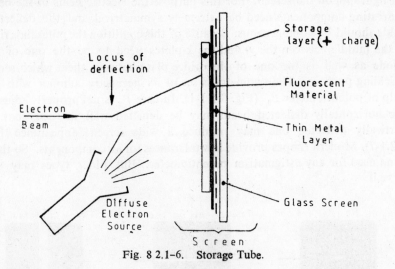

Fig. 8 2.1-6.   Storage Tube.

marks, a neutral trace is left behind amidst the positively charged area of the storage layer. This happens within the 'writing' mode of the scope which stores the information this way. For watching the trace a second source needs to be employed which emits electrons in a diffused way with comparatively low kinetic energy. These electrons (uniformly distributed) are sprayed at the storage layer. At the location of the neutral trace they may pass and can excite the fluorescent layer for light emission. The trace becomes visible. At the other positively charged area the electrons cannot pass. Their negative charge recombines with the positive charges and this part of the field remains in dark. At first a bright trace (or spot) appears within a dark surrounding producing a beautiful contrast. But gradually, more incoming electrons discharge the positive area of the surrounding field and finally they all manage to pass over to the screen thus exciting light emission everywhere. The contrast emerges and the stored information disappears gradually, because the whole area gets neutralized eventually. This happens within the scope mode of 'view'. Usually a short time of viewing may be sufficient to assess the measurand. So one should switch over to the mode 'store' in order to sustain the information. The storage time ranges up to 60 hours for storage scopes these days, even for the switched off device. Advances in the technology of insulation materials provide this feature. In case a new trace should be stored, the whole layer needs be 'erased' first. This is effected by recharging its entire area positively.

Storage tubes are expensive. Successful efforts were undertaken, therefore to store the information in digital memories (see chapter 8.2.4).

Astigmatism

The cross section area of the beam should be a narrow circle, and so should the light spot on the screen. For this purpose the electric fields of the beam generating components need be extremely symmetrical and the deflection fields should be homogeneous. Because of this condition the potentials right in the middle between the y- and the x-plates need to be the one of the anode as well as the one of the orifice plate between them which serves shielding purposes. The symmetry is set to its necessary amount with the help of potentiometer P_3 (Fig. 8.2.1-1). In case P_3 is not properly adjusted the horizontally deflected beam may be depicted well in focus but the vertically deflected one may produce a wide screen appearance (Fig. 8.2.1-7). Modern scopes provide low tolerances for all components. So there is no need for any astigmatism potentiometer. But older types may still have it.

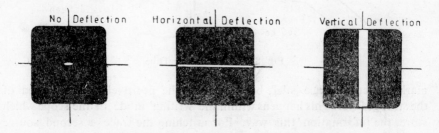

Fig. 8.2.1–7. Effect of Astigmatism.

8.2.2 Time Base

In order to deflect the beam horizontally proportional to time a voltage U_X is needed which rises linearly with time. It is used to control the position of the light spot on the screen in x-direction. Its lowest value $U_{X\,min}$ should set the beam to the left edge of the screen and its highest one $U_{X\,max}$ to the right edge. The whole run should be provided repeatedly. The switching back from the right to the left edge should be effected as fast as possible. So a saw tooth x-deflection voltage serves our purpose (Fig. 8.2.2–1). For trace 2 the deflection is slower than for trace 1. Increasing the slope speeds up the deflection which allows to adopt the depiction to high frequencies of the measurand.

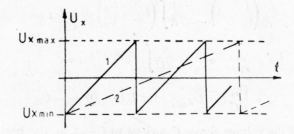

Fig. 8.2.2–1. Voltage Across the x-Plates for Fast (1) and Slow (2) Deflection.

Miller-Integrator

In case a capacitor is fed by a constant current i its voltage will increase linearly. This is provided by the MILLER integrator circuit employing an operational amplifier (Fig. 8.2.2–2).

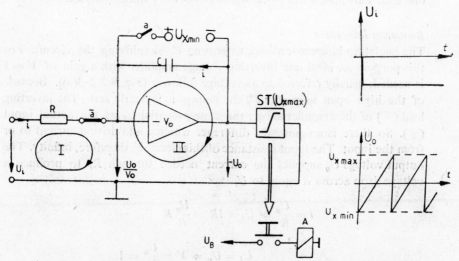

Fig. 8.2.2–2. Principle of Saw-tooth Generation Using the MILLER-Integrator.

Because of the very high open loop gain V_0 of the amplifying element, nearly no current is needed to feed the amplifier. All the input current is determined by R and compensated by the one through C. Using loop I and loop II, the function of the output voltage U_0 with respect to the input voltage U_i may be obtained:

$$\text{I:} \qquad U_i = R_i + \frac{U_0}{V_0} \quad \Rightarrow \quad i = \frac{1}{R}\left(U_i - \frac{U_0}{V_0}\right)$$

$$\text{II:} \quad + \frac{U_0}{V_0} = \frac{-1}{C}\int i\, dt + U_{X\,min} - U_0$$

$$U_0\left(\frac{1}{V_0} + 1\right) = \frac{-1}{CR}\int\left(U_i - \frac{U_0}{V_0}\right) dt + U_{X\,min}$$

$$U_0' = \lim_{V_0 \to \infty} U_0 = -\frac{1}{RC}\int U_i\, dt + U_{X\,min}$$

$$= \frac{U_{i0}}{RC}t + U_{X_{min}} \quad \text{for } U_i(t = 0) = -U_{i0}$$

In case the voltage across C was set to $U_{X\,min}$ at the beginning $(t = 0)$ the output voltage U_0 rises positively for a negative constant input voltage U_i. A SMITH-trigger ST may sense $U_{X\,max}$ and instantly set back the voltage across C to $U_{X\,min}$. As soon as this happens via the contact a of the A-relay the other contact $\bar{a}$ breaks the input voltage U_i off the integrator input. A relay was used only for easy explanation of the action principle. Of course modern scopes provide this feature by employing electronic means.

It is a disadvantage of the Miller integrator that C is not grounded at one end. This makes the switching over circuitry quite complex.

Bootstrap Integrator

The bootstrap integrator allows to ground C, simplifying the circuit. For this purpose an ideal non inverting voltage amplifier with a gain of $V = 1$ is needed, usually referred to as voltage follower [Fig. 8.2.2–3(a)]. Because of the high open loop gain V_0 the voltage u_ε is nearly zero. The inverting lead $(-)$ of the amplifier shows the same potential U_i as the actual input $(+)$. So there is no potential difference which could drive a current to or from the input. The input resistance of this circuit is, therefore, infinity. The output voltage U_0 supplies the current needed through R, to produce a voltage drop across R equal to U_i

$$i = \frac{U_0}{R} \quad \Rightarrow \quad U_i = iR = \frac{U_0}{R}R$$

$$U_i = U_0 \quad \Rightarrow \quad V = \frac{U_0}{U_i} = 1$$

a)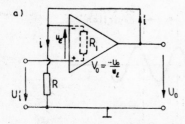

b)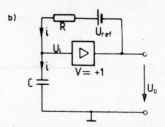

Fig 8.2.2-3. Voltage Follower (a) and its Use as Part of the Bootstrap Integrator (b).

Such a voltage follower is used to assemble the bootstrap integrator, [Fig. 8.2.2-3 (b)]. As $U_0 = U_i$ the current $i = U_{ref}/R$. Because of the very high input resistance, i cannot even partly go into the amplifier but continues to pass through the capacitor C. The reference voltage U_{ref} and the resistor R are constant. and therefore i is constant, too. So C is charged with a constant current i and its voltage which is U_i rises linearly with time. Due to $V = 1$, U_0 follows directly, as shown with the help of the following equations:

$$U_c = U_i = U_0 = \frac{1}{C} \int i dt + C_1$$

$$= \frac{1}{RC} \int U_{ref}\, dt + C_1$$

$$= \underbrace{\frac{U_{ref}}{RC} \cdot t}_{m} + C_1$$

The bootstrap integrator is superior compared to other integrators for assembling a saw tooth generator. This is because of its capacitor C is grounded, thus allowing simple electronic means to effect a fast switching from $U_{x\,max}$ to $U_{x\,min}$.

8.2.3 Time Depiction of Time Dependent Voltages

The slope m of the saw tooth can be changed preferably with the help of C and R. For setting the range to a proper value, C may be changed with the help of a selector switch. R can be used for fine adjustment. This way any time expansion may be arranged. In order to obtain a stable depiction of the measured voltage, the frequency of the measurand needs to be an integer multiple of the saw tooth frequency. If this condition is met the electron ray will trace the same line for each repeated x-deflection. The depiction will appear as stationary graph.

The synchronization and the trigger method provide means to match the saw tooth frequency to the one of the measurand.

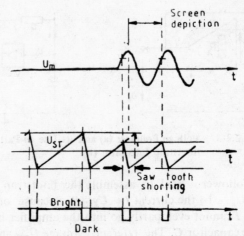

Fig. 8.2.3-1. Voltages and Screen Depiction of a Synchronized Scope.

Synchronization

Employing the synchronization method, the saw tooth frequency is matched by shortening its period duration. This is provided by the measured voltage U_m, (Fig. 8.2.3-1). For this purpose a voltage level U_L is sensed which is composed from the saw tooth voltage U_{ST} and the measured voltage U_m:

$$U_L = U_{ST} + U_m$$

At the instant of reaching U_L the fly back of the beam is immediately effected. The saw tooth is shortened unfortunately. So the trace cannot reach the right edge completely. Of course it needs some time to get the beam back to the left edge of the screen. In order to avoid a visible return trace, the depiction is manipulated "dark" with the help of the Wehnelt cylinder, which is supplied with a high negative voltage during this time. This way the flow of electrons is interrupted. During the time which is needed for the fly back, no information about the measurand is obtainable. This is certainly a disadvantage. The deflection time of the beam in *x*-direction needs to be at least one period duration of the measurand or integer multiples of it. Consequently only one or several periods of the measurand can be depicted, but never parts of the period duration. This is another disadvantage.

Synchronization can be implemented with the help of an "extern" voltage or the measurand itself which is labelled as "intern" at the related switch or by the "line" voltage. These three synchronization modes are available.

Trigger Method

The trigger method avoids both disadvantages of synchronization and is therefore far superior. Due to the decline of prices of electronic equipment, even low cost oscilloscopes provide triggered time bases.

The trigger method allows a standing depiction even if the depicted number of period durations is not a whole multiple of the rise time of the saw tooth. But its frequency is of course matched to the one of the measurand (Fig. 8.2 3-2), by implementing a "waiting time". This way even parts of the period duration of the measured voltage can be depicted.

For this purpose the trigger instant of the saw tooth voltage is linked to a certain instantaneous value of the measurand, the trigger level, and its slope (for recognizing the polarity). Once the trigger level is reached by U_m at a positive slope (as it is depicted), the beam starts at the left edge, for at that instant $U_{ST} = U_{X\,min}$. The light spot travels with a certain speed in x-direction across the screen according to the slope of U_{ST} and finally reaches the right edge. U_{ST} is equal to $U_{X\,max}$ then. Immediately after $U_{X\,max}$ was reached the beam is switched back to the left edge where the spot "waits" to be released another time. The screen depiction shows a trace of the measurand during the positive slope of U_{ST}. No depiction appears during the fly back time and the waiting time of the spot.

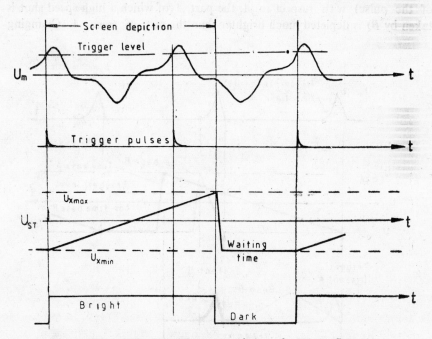

Fig. 8.2.3-2. Voltage and Screen Depiction of a Trigger Scope

The trigger "level" may be shifted throughout the whole voltage range between peak to peak value of the measurand. Doing this the depiction appears to be shifted in x-direction.

The slope of the saw tooth voltage U_{ST} can be changed in wide ranges from very slow to very fast, allowing to depict many periods of the measurand on the screen but also only parts of it.

Trigger Scope with Double Time Base

Sometimes it is quite feasible to depict a measurand with several period durations, to allow a good general overview and at the same time to depict a certain small time interval only to facilitate a close detailed time investigation. A scope with two time bases allows to do so. It is feasible if the measurand consists of short pulses followed by long pauses. For the Fig. 8.2.3-3, a double beam oscilloscope was assumed which provides two independent beams.

The first one A obeys the control of time base A. As explained in the previous passage, the depiction of trace A is obtained. As the slope of $U_{ST, A}$ is slow enough, about one and a half period durations of U_m are depicted. The trigger level of time base B is related to the saw tooth voltage of time base A. Changing this level causes beam B to be triggered at different time instances. Advantageously the starting instant of beam B in combination with a much faster slope was selected this way to depict the impulse length of U_m over the full x-range of the screen.

For purposes of easy location of trace B (which is a slow motion picture of the pulse) with respect to A, the part A (of which a high speed shot is taken by B) is depicted much brighter than the rest of trace A. Changing

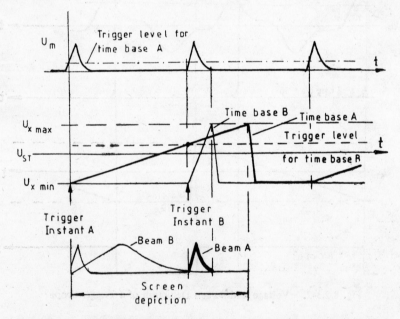

Fig. 8.2.3-3.   Voltages and Screen Depiction of a Trigger Scope having two Time Bases.

the trigger level for beam B causes the brighter part of trace A to "creep" over the peaks and valleys of the measurand U_m like a caterpillar.

Sampling Oscilloscope

Measurands of very high frequencies (> 100 MHz) cannot be directly depicted using the trigger method. But the sampling method can cope up with such signals in case they are periodical. The measured voltage U_m is continuously scanned by taking a sample, during each cycle. But the time instant of measurement is slightly shifted consecutively for each new sample.

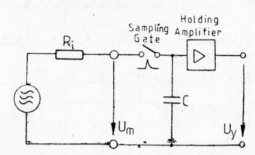

Fig. 8.2.3-4. **Principle of the Input Circuit of a Sampling Scope.**

The sampling gate is periodically activated for a very short sample time, (Fig. 8.2.3-4), scanning the voltage U_m which is provided by its source generator having the internal resistance R_i. During the sample time the capacitor C is charged. It takes over a voltage equal to the instantaneous value of U_m during the sample time in case the time constant $\tau = R_i \cdot C$ is small enough. After the sampling gate has opened, the charge of the capacitor remains unchanged and so does its voltage. It is held constant until the gate is switched on again and another U_m may change the capacitor voltage. This means: during the pause between two sample pulses, the capacitor voltage stays constant and an amplifier provides it at a higher (energy) level as U_y. The detailed voltages are given in Fig. 8.2.3-5 as functions of time, which allow to grasp the matter easily.

The rising signal U_m to be measured causes a trigger pulse periodically. The command for the sampling gate is given after a time delay. This delay is arranged to be Δt after the first trigger pulse, $2\Delta t$ after the second, $3\Delta t$ after the third, etc. Each sample pulse connects the capacitor C for a very short time to the measurand and takes over its instantaneous voltage. It is held during the sample breaks and provided by the holding amplifier as U_y to control the y-plates of the scope. U_y changes in steps following U_m. If there are 10 sample pulses during 10 period durations of U_m, the voltage U_y has followed in steps only once. The x-plates are supplied by a step voltage arising in equal steps from $U_{x\,min}$ to $U_{x\,max}$. (The generation of such a voltage U_{ST} is described in chapter 12.3, see fig. 12.3-3). In this way the beam is deflected in steps from the left edge to the right edge. The step width is 1/10 of ($U_{x\,max}$ $U_{x\,min}$). In this way the light emitting spot on the screen leaps from one depicted dot to the other. A dotted trace occurs. It gives quite a good imagination of the actual signal. The density of the dots may be increased considerably by choosing a shorter sample delay time Δt. In this way the scanning is less coarse providing a depiction which may be considered to be quite a steady line. But the time for tracing one period

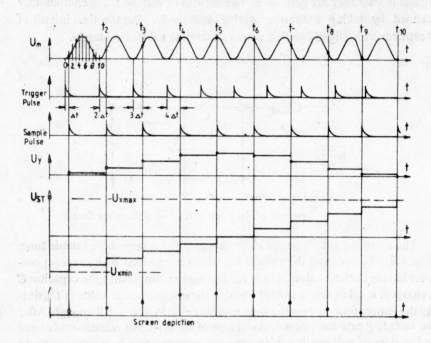

Fig. 8.2.3-5. Voltages and Screen Depiction of a Sampling Oscilloscope.

duration increases. The number of sampling pulses needed for a full depiction of one cycle, times the period duration of the measurand, equals the time for depicting the trace once. The screen shows a time expanded appearance of U_m. It is a stroboscopic presentation of the measurand.

It should be pointed out that a sampling device can also be effectively used for 50 Hz measurands to be investigated with slow plotters. They usually cannot follow directly, but employing this method the speed may be taken down by a factor of 10 or 100 to which the plotter can respond. Normally the step function U_y would be drawn graphically. But one can also use the pen lift in order to obtain a dotted curve by commanding it accordingly.

Sample time and sample rate are calculated in chapter 13.4 "Sample (Aperture) Error".

8.2.4 Storage Oscilloscope with Digital Memory

To store the instantaneous value of a measurand is certainly feasible, if it is non repetitive. Traditional storage scopes are expensive and their handling is quite complex. Digital storage oscilloscopes avoid both disadvantages and provide additional features: higher accuracy, unlimited storage time, (in combination with magnetic outer stores, like floppy discs) unlimited storage capacity, easy comparison of stored and live signals, and analog signal output for slow plotters. A cursor allows to address a certain spot and a numerical output of the spot coordinates is digitally displayed. An instrument having these features is provided by NICOLET within its Explorer-series.

Fast analog to digital coverters are the key devices for digital storage oscilloscopes. They instantaneously scan the whole pulse to be investigated. A typical resolution of 50 μV for each time instant of scanning is obtainable. The time base operates with a typical time error of 0.02 % only.

Signals down to 0.2 μs of pulse duration can be handled. The measured voltage is converted into digital words of 12 bits length which are stored into integrated memories. A continuous program routine reads out the memory contents repetitively with a free selectable speed. The digital values are converted into analog voltages, which are displayed. The evaluation is facilitated by employing an eletronic cross web with the help of which any spot on the screen can be addressed. This feasibility is called a "cursor". Its y-voltage and x-time coordinates are digitally displayed within the screen as decimal numbers. The near region around the cursor can be magnified up to 64 times presenting minute details of the measurand.

This latest scope technology allows to connect digital computers, broadening the application range of scopes considerably.

8.3 LIMITING FREQUENCY F_0 AND BANDWIDTH

Deriving the equation for the deflection S we assumed constant field strength E between the deflection plates as long as the electrons pass them. But as oscilloscopes are particularly designed for AC measurements, this assumption of constant E can be fulfilled up to a certain degree only. The permissible indication error ε determines the highest possible frequency, the limiting frequency f_0, of the voltage U_y to be measured ($U_y = E \cdot d$).

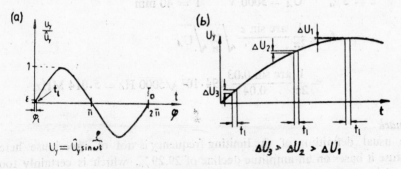

Fig. 8.3-1. Terms for Calculating the Limiting Frequency of a Scope.
(a) Full Trace of sin ωt.   (b) Detail Around Zero.

The sine determines the change of U_y. Let us allow a relative change

$$\varepsilon = \Delta U/U = U_y/\hat{U}_y$$

of a few per cent during the passing time t_l. The shortest period duration $T_0 = 1/f_0$ gives the fastest change at its zero crossing.

$$\varepsilon = \frac{U_y}{\hat{U}_y} = \sin \underbrace{\omega_0 t_l}_{\varphi_l}$$

$$\omega_0 = 2\pi f_0$$

t_l is the time which an electron needs to pass the plates of the length l. Meanwhile the argument of the sine increases by φ_l. From the x-axis of the graph the following relation may directly be read as

$$\frac{T_0}{t_l} = \frac{2\pi}{\varphi_l}$$

$$f_0 = \frac{1}{2\pi} \frac{\varphi_l}{t_l}$$

with

$$\varphi_l = \text{arc sin } \varepsilon$$

f_0 will be

$$f_0 = \frac{1}{2\pi} \frac{\text{arc sin } \varepsilon}{t_l}$$

The passing time t_l is determined by the length l of the deflection plates and the velocity v_A of the passing electrons: $t_l = 1/v_A$. v_A depends on the anode voltage U_A only, which accelerates the electrons as shown before:

$$v_A = \sqrt{2eU_A/m}.$$

Using t_l and v_A, f_0 can be calculated from known quantities:

Ex. $\varepsilon = 3\%$ $U_A = 5000 \text{ V}$ $1 = 40 \text{ mm}$

$$f_0 = \frac{1}{2\pi} \frac{\text{arc sin } \varepsilon}{l} \sqrt{\frac{2e}{m}} \sqrt{U_A}$$

$$= \frac{1}{2\pi} \frac{\text{arc sin } 0.03}{0.04} 594 \cdot 10^3 \sqrt{5000} \text{ Hz} = 5.014 \text{ MHz}.$$

Remark

The usual definition of the limiting frequency is not of good use here because it bases on an amplitue decline of 29.29%, which is certainly too large for measurement purposes. In this case the limiting frequency would be calculated as 49.673 MHz which is unrealistic due to the high indication error. However the user of an oscilloscope should be aware that manufacturers occasionally give the limiting frequency as such (3dB-decline of the amplitude) by labeling the bandwidth *B*.

Relation between Bandwidth and Rise Time:
The input of the vertical channel may be considered as an **RC**-low pass (LP), see Fig. 8.3-2.

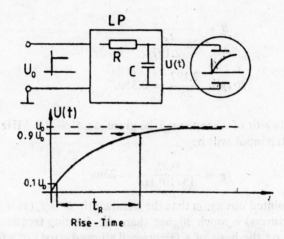

Fig. 8.3-2. Vertical Input of a Scope (Output Response $U(t)$
to an Input Step U_0.

For an input step U_0 at the time instant $t = 0$ the response voltage across the y-plates is an exponential function with the time constant

$$\tau = R \cdot C:$$

$$U(t) = U_0 \left(1 - e^{-t/\tau}\right)$$

The time which the voltage $U(t)$ needs to rise from 10% to 90% of its final value is defined as the risetime t_R. It is a typical quantity for a scope and can be determined from the previous equation:

$$t = \tau \ln \frac{1}{1 - \dfrac{U(t)}{U_0}}$$

$$t_R = \tau \left(\ln \frac{1}{1 - 0.9} - \ln \frac{1}{1 - 0.1}\right)$$

$$= 2.197\tau \quad (*)$$

Let us define the circular cut-off frequency as $\omega_c = 1/\tau$. Using this definition (which is different to the one used for the limiting frequency f_0 of the previous paragraph) the bandwidth B can be obtained and also the product of $B \cdot t_R$ which is a constant. This is an important result: A scope providing a wide band width can depict pulses of short rise time.

$$f_c = \frac{1}{2\pi \cdot \tau} = B$$

Using equation (∗) the bandwith B can be determined by measuring the risetime t_R.

$$B = \frac{1}{2\pi \dfrac{t_R}{2.197}}$$

$$B \cdot t_R = \frac{2.197}{2\pi} = 0.350.$$

Ex.   The bandwidth of a scope may be given as $B = 15$ MHz. The rise-time t_R for a step input will be

$$t_R = \frac{0.35}{15 \cdot 10^6 \text{ Hz}} = 23\text{ns}$$

It should be pointed out again that the bandwidth $B = f_c$ (as it is stated by scope manufacturers) is much higher than the limiting frequency f_0 as it was calculated on the basis of a fairly small allowed error ε of a few % only. But for a measurement frequency $f = f_c = B$, the indication error would amount to 29.29%, as previously stated.

If a step signal is applied, having its own rise time of t_{R1}, a total rise-time $t_{R,\text{tot}}$ is displayed:

$$t_{R,\text{tot}} = \sqrt{t_R^2 + t_{R1}^2}$$

The actual rise time t_R of the scope can be calculated from this formula. This presumes a linear phase φ with respect to the circular frequency:

$$\varphi = f(\omega) = a \cdot \omega.$$

8.4   LISSAJOUS FIGURES

Lissajous figures are depictions of two voltages applied to the y- and x-plates which are periodical time functions. But the depiction on the screen is independent of the time. Whether the frequencies of U_x and U_y change fast or slow is not decisive. The trace will be the same in any case.

Frequency Measurements

Lissajous figures can be obtained from independent voltages U_x and U_y. In this case they are usually used for frequency measurements. The unknown frequency can be determined if the other one is known. Let us consider Fig. 8.4–1. The first graph depicts a voltage U_x of frequency f_x and a voltage U_y

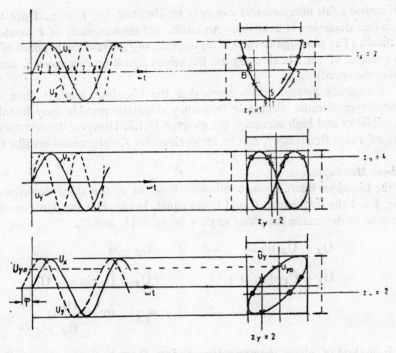

Fig. 8.4-1. Lissajous Figures.

of the frequency $f_y = 2f_x$. The Lissajous figure can be sketched with the help of the voltage pairs U_x, U_y for the 9 sample points. One maximum distance in x-direction (n_x) from 7 to 3 is found and two maximum distances in y direction (n_y) from 9 to 7 and from 1 to 3. Obviously the frequency f_y can be determined from

$$f_Y = \frac{n_Y}{n_X} f_X = \frac{2}{1} f_X = 2f_X = \frac{1/z_y}{1/z_x} f_x = \frac{1/1}{1/2} f_x = 2f_x = \ldots$$

The same result for the frequency f_y is obtained from the second depiction, though the Lissajous figure appears quite different due to the phase shifted U_y. This formula of f_y proves to be true for the third depiction as well: $n_x = 1$ and $n_y = 1$, so $f_x = f_y$. —If one frequency is known the other can be determined. But the ratio of f_y/f_x needs to be a proper fraction in order to obtain a standing screen depiction. This condition can be met with the help of a tunable frequency generator for f_x. As long as the frequency condition is not fulfilled ($f_y \neq f_x \, n_y/n_x$) the depiction changes its shape continuously. The appearance moves with a period of time T for one cycle. The unknown frequency f_y depends on f_x in the following way

$$f_Y = f_X \frac{n_Y}{n_X} \left(1 \pm \frac{1}{T \cdot f_X} \right)$$

Of course a fair measurement can only be obtained for $T = \infty$. Only then a stable depiction is available. An additional measurement of T would be difficult. (The plus sign is valid if an increase of f_x decreases the speed of the movement of the Lissajous figure, the minus sign if an increasing f_x accelerates the speed).

Frequency measurements employing the Lissajous method have lost importance because electronic frequency counters provide easy handling possibilities and high accuracy, see chapter 13.1.2. However the accuracy of the reference frequency f_x can be maintained for f_y-determinations like this.

Phase Measurements

If the Lissajous figure is one continuous loop as in the third depiction of Fig. 8.4–1 the frequencies f_x and f_y are equal. In this case the figure can also be used to determine the phase angle φ between U_x and U_y.

$$U_X = \hat{U}_X \sin \omega t \qquad \underline{\omega t = 0} \qquad U_X = 0$$

$$U_Y = \hat{U}_Y \sin (\omega t + \varphi) \qquad\qquad U_Y = \hat{U}_Y \sin \varphi = U_{Y0}$$

$$\Rightarrow \varphi = \text{arc} \sin \frac{U_{Y0}}{\hat{U}_Y}.$$

This method of phase determination suffers from bad errors especially if the argument of the arc sin is near one. This means φ is near 90°. But in case the readings U_{y0} and $\hat{U}_y$ can be obtained with high accuracy (at least for φ somewhat apart from 90°) a decent low error in φ may be obtained. This is possible if a digital storage scope is employed, see chapter 8.2.4. The use of the cursor and the digital reading possibility allow a 50μV resolution of the measured voltages, which is sufficiently good if U_x and U_y are large enough.

9 Disturbances of Measurement Data
(Shielding, Grounding, Guarding)

Transmitting measurement data, subjects them normally to external disturbances due to capacitive and inductive stray fields of nearby high voltage and heavy current lines. (Fig. 9-1). Even a resistive coupling may occur, due to insufficient insulation between the power line and the measuring system. Other long term disturbances like thermoelectric or galvanic voltages may be superimposed additionally to the measurand U_M.

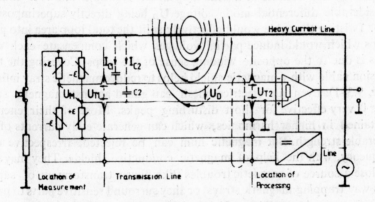

Fig. 9-1. Disturbances of a Measuring System.

Wrong grounding may cause other problems. There are many ways to avoid, minimise, or eliminate disturbances effectively.

9.1 CAPACITIVE DISTURBANCES (Common Mode Type)

Capacitive disturbances occur through capacitive coupling of power lines with transmission lines of measuring systems. The currents I_{C_1} and I_{C_2} may cause a voltage drop across the internal resistance of the measuring transducer, thus superimposing a hum on the measurand U_M. As soon as both currents are equal, they compensate each other. But normally they are different and the question arises how to avoid their effect on U_M. Very effective results can be obtained by using twisted cables which make the coupling capacitances C_1 and C_2 equal to each other. (Fig. 9.3-1). This way the upper and the lower wire of the transmission line form alternatively partial capacitors from loop to loop of the twists. They add up to the total capacitance C which becomes equal to C_1 and to C_2 of the total length of the cable. An additional electric shield may complement by reducing the coupling effectively. In this case the currents I_{C_1} and I_{C_2} would mainly be guided to ground. They are altogether small for a good distance between power and transmission line. Both lines should not be near to each other and never be contained inside the same cable.

These common mode disturbances are signified by equal phases and (nearly) equal amounts of their voltages being applied to the inputs of the amplifier. They are rejected, with differential amplifiers, from further processing (chapter 10.3.1). Of course, they can be filtered away, too. However, a filter may extend the response time of the system unacceptably.

9.2   MAGNETIC DISTURBANCES (Differential Mode Type)

Magnetic fields of heavy current lines may impose considerable hum on the measurand U_M, in case the measuring line transmitting U_M catches quite some stray flux because of a wide loop area. It would allow to induce a considerable differential mode voltage U_D being directly superimposed on U_M. Twisting the cable would effectively split the total loop area into partial loops which would induce partial voltages which compensate each other. This is due to the opposite winding sense of the loops. Covering the transmission cable with a magnetic shield helps to reduce magnetic stray influence (Fig. 9.3–1). Even an electrical mesh can guard against magnetic strays. This is very effective for short disturbing peaks. Most of their energy is contained in higher harmonics, which can generate eddy currents of considerable strength. But magnetic hum can be rejected, irrespective of its frequency, with the help of magnetic conductive shields. They may either enclose a source of magnetic troubles (like power transformers of supplies this way trapping magnetic strays) or they surround sensitive parts to protect them from outside. —Differential mode disturbances are signified by opposite phases (but equal amounts) of their voltages being applied to the inputs of the amplifier. They may be rejected by employing an integration of the measurand which is executed over some full periods of the disturbance, see chapters 11.2.2 to 11.4.

9.3   ATTENUATION OF DISTURBANCES BY SHIELDING AND TWISTING OF THE INPUT CABLE

Electrical disturbances due to capacitive coupling are common mode interferences. They effect both inputs quite the same and are therefore rejected, see next chapter. But due to stray fields of power cables, magnetic disturbances also occur. They induce, within the input loop, a voltage U_D which cannot be distinguished from the signal to be amplified. One can partly cope with these difficulties using a twisted input cable. This makes the effective loops short. If furthermore the wire conductors are close to each other (due to a thin insulation) the area of each loop is small, therefore cannot catch considerable magnetic stray flux. Consequently the induced voltage is small. A magnetic shield yields the same effect.

Fig. 9.3-1 shows the importance of shielding and twisting with respect to the disturbing induced voltage U_D. All cables are short-circuited at one end. The first one serves as a reference showing for the other types how often U_D will be increased or decreased. The comparison is impressive.

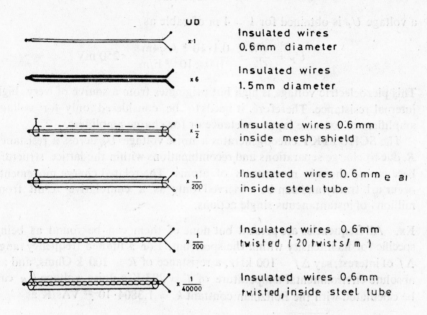

Fig. 9.3-1.   Comparison of the Attenuation Effect depending on the Kind of Input Cable and its Shielding (SIEMENS Report)

9.4   LONG TERM DISTURBANCES

Thermoelectric voltages may occur at any connection of different metals (Fig. 9-1). Two junctions at different temperatures are a thermo couple which generates a voltage U_T (SEEBECK effect) being proportional to their temperature difference (chapter 17.7.1). The easiest way to get rid of U_T is to provide equal temperature at any two connection points which could act as a thermo-couple. Some ten micro volts may frequently be observed if no preventive care is taken. They originate from a source of low internal resistance and could give rise to considerable undesired currents.

The voltage of a GALVANIC element with some ten milli volts or even more than $U_G = 100$ mV may seriously interfere with a measurement. Humidity, together with contaminants of the air, may act as a perfect electrolyte between different metals not being sufficiently covered by waterproof insulation materials. Electrolysis of this kind corrodes the metals concerned. It has the tendency of causing failures and leaving the equipment faulty soon and, therefore, should be taken care of in design.

Ex. Bending a cable, a charge generation of about $Q = 0.1$ nC may occur, due to the piezoelectric effect which derives from the mechanical stress towards the dielectric insulation between two wires. With a cable capacity per meter length of

$$C' = C/l = 100 p \text{ F/m}$$

a voltage U_P is obtained for $1 = 4$ m of cable as

$$U_P = \frac{Q/l}{C'} = \frac{0.1 \cdot 10^{-9} \text{ As/4m}}{100 \times 10^{-12} \text{ F/m}} = 250 \text{ mV}$$

This piezoelectric voltage is high but originates from a source of very high internal resistance. Therefore, it needs to be considered only for voltage amplifiers of extreme input resistance or for charge amplifiers.

The SCHOTTKY effect generates a noise voltage U_N across a resistance R, due to charge separations and recombinations within the lattice structure because of thermic movements of atoms The related charge movements occur arbitrarily and can be observed only as a cumulative result from millions of instantaneous single actions.

Ex. All frequencies are present but none of them can be found as being specific for its intensity within the spectrum. For a limited frequency range Δf of interest, say $\Delta f = 100$ kHz, a resistance of $R = 100$ k Ohms, and an absolute environmental temperature of $T = 293$ K a noise voltage U_N can be calculated with the Boltzman constant $k = 1.3804 \cdot 10^{-23}$ VAs/K as

$$U_N = \sqrt{4kTR\Delta f}$$

$$= \sqrt{4 \cdot 1.3804 \cdot 10^{-23} \frac{\text{VAs}}{\text{K}} \, 293 \text{ K} \cdot 10^5 \frac{\text{V}}{\text{A}} \, 10^5 \frac{1}{\text{s}}} = 545 \text{ } \mu\text{V}$$

Whether U_N may interfere, depends on the impedance of the following amplifier loading the noise source which provides the internal resistance R.

9.5 LOOP DISTURBANCES

9.5.1 Coaxial Cable Loop

Normally measuring devices are connected to each other by means of coaxial cables. As the instruments are usually supplied from the same line, a current loop is set up tending to catch magnetic field disturbances which may induce a considerable current I. This way a voltage drop is caused across the shield resistance R_s of the coaxial cable (Fig. 9.5.1-1). ΔU_s adds directly to the measurand. Of course, the same principle is set into action with ordinary connection wires between the devices. By supplying one of them with a line separating transformer the current circle for I is interrupted, this way eliminating this disturbance completely.

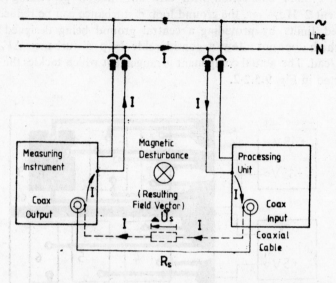

Fig. 9.5.1-1.   Loop Disturbance ΔU_s of the Coaxial Shield due to Instrument Connection with Coaxial Cable.

In case the current I can also cause a voltage drop across the neutral conductor N (due to a long distance between the devices) this adds in the same way as another ΔU_s to the measurand. This case is separately dealt with in the next chapter.

9.5.2 Ground and Supply Loop

Ground loops are made up by wrong connections of the supply voltage to different components of the complete circuit (Fig. 9.5.2-1). If the supply current I of the component 2 is high it causes a voltage drop ΔV_G directly

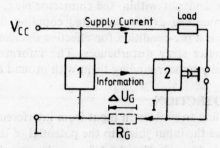

Fig. 9.5.2-1.   Loop Disturbance ΔU_G due to a Ground Loop.

adding up to the information. This aggravates in case device 2 switches a heavy load to the supply. Such loads should, therefore, never share a common ground connection with any information processing unit like component 2. However, the ground loop disturbance can be reduced below permitted limits by providing a central ground being designed as a star point which does not allow a considerable loop disturbance ΔV_G of the ground lead. The actual component arrangement which tackles the problem is depicted in Fig. 9.5.2-2.

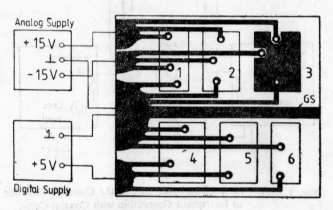

Fig. 9.5.2-2.   Supply Scheme and Component Arrangement avoiding Ground Loops.

Analog elements 1 to 3 and digital units 4 to 6 are arranged on the same printed card but carefully separated from each other which is supported by a ground separating area *GS* not permitting direct electric coupling from the upper to the lower side and vice versa.

The analog element 3 is additionally shielded. Being a high frequency device this measure may be needed. At least for the digital elements, placing fast tantalum capacitors as near as possible to their supply leads, can help to keep the supply stable. The ground of the analog and the digital supply voltage should be made only at one point right on the card with neighbouring connector pins and not within the connector plug, let alone near the supplies. Proper grounding avoids unwanted coupling of different circuit stages internally. It is a pre-condition for effective shielding of the information line against outer stray disturbances. The information lines may be protected additionally by surrounding them with ground areas.

9.6   GUARD PROTECTION

An alternative method to protect an input from interferences is the principle of guarding. It takes the input shield to the potential of the guarded part of the circuit, which is shown in Fig. 9.6-1 for a voltage and a current input. The guard of depiction (a) is on input potential U_i because the operational amplifier makes its inputs potentially equal (refer also to Fig. 8.2.2-3). So there is no potential difference between the guard and the center conductor of the coaxial input. This way a disturbing voltage which could be caught by the guard only is shorted with respect of the input and cannot effect U_I.

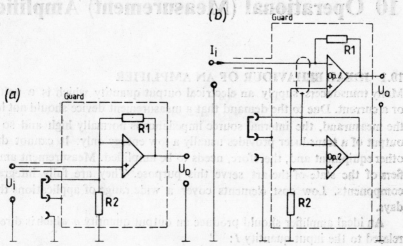

Fig. 9.6-1.   Guarded Voltage (a) and Current Input (b).

The guard of circuit (b) is connected to a voltage follower (Op. 2) allowing current measurements. The guard can include a shielding mesh enclosure for the internal input leads of the amplifier (Op. 1).

Devices being equipped with a guard terminal, usually provide the feature which allows to switch over the input shield from guard potential to ground.

10 Operational (Measurement) Amplifier

10.1 IDEAL BEHAVIOUR OF AN AMPLIFIER

Many transducers supply an electrical output quantity which is a voltage or a current. Due to the demand that a measurement device should not load the measurand, the internal source impedance is normally high and so the output of a transducer provides usually a low energy only. It cannot drive other equipment and, therefore, needs to be amplified. Measurement amplifiers of the state-of-the-art serve this purpose. They are fully integrated components. Low cost elements cover a wide range of applications these days.

An ideal amplifier should produce an output quantity o which is directly related to the input quantity I:

$$o = V \cdot I$$

V is the gain of the amplifier.

10.2 ADDITIVE AND MULTIPLICATIVE DISTURBANCES

The amplifier is usually subject to outer disturbances, which change the ideal transfer equation. They may introduce additive and multiplicative terms.

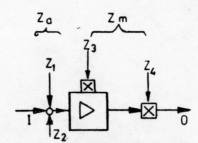

Fig. 10.2-1.   Amplifier and its Disturbing
Variables.

As shown in Fig. 10.2-1, the input may be affected by input disturbances Z_1, such as inductive stray fields or thermo electric voltages, etc. A Q-point shift of the first amplifier stage may also add disturbances Z_2. They usually affect the input signal I considerably, because it is still at a low energy level. All input disturbances work additively and may be signified by the symbol Z_a. They are amplified together with I.

The single amplifier stages are predominantly influenced by surrounding disturbances Z_3 which change the features of their components. They are temperature, humidity, age and supply changes. They all affect the gain V by a multiplying factor. (Of course the intermediate amplifier stages are also effected by additive disturbances. But as the signal of interest is amplified to a high energy level by then, it cannot be changed considerably).

The output feeds the load which might change. Also the resistance of the output cable may suffer from changes due to outer disturbances Z_4. They affect the output quantity o mutiplicatively just as the disturbances Z_3 do. They are, therefore, referred to as Z_m.

All disturbances cause a change in the output quantity, which does not only depend on the input quantity, as it should, but also becomes a function of V, Z_a and Z_m in reality :

$$O' = \underbrace{(I + Z_a)}_{I'} \times \underbrace{V \cdot Z_m}_{V'}$$

10.3 ELIMINATION OF DISTURBANCES

By employing appropriate means, most of the disturbances may be effectively rejected. Simplest measures are-good shielding of the input cable and the use of low-noise input-components for the first amplifier stage. But more important is to employ a differential amplifier to get rid of additive disturbances and to make use of negative feed back circuitry to cope up with multiplying terms.

10.3.1 Differential Amplifier

Integrated Circuit

The differential amplifier employs two input stages which work opposite to each other. Manufacturers of integrated circuits were successful in supplying amplifiers with precisely equal input stages. Due to the fact that they are manufactured through the same process, both stages provide equal gain and equal temperature coefficient of their gain. Being symmetrically positioned on the same chip, their components will be subject to the same temperature. So if there should be a drift in one stage the other will show the same drift. Operational amplifiers are usually multi-stage devices, to obtain a high gain. But for simplicity reasons, a symmetrical difference amplifier, with two double stages is shown in Fig. 10.3.1-1.

All transistors are fed from a positive supply via their collector resistance R_C. For short-circuited input leads ($U_I = 0$) the collectors of the outer stages are at equal (high) potential. Consequently the inner stages are equally fed and their collectors provide the same (low) potential, producing an output voltage $U_0 = 0$. For a DC input voltage U_I, with the indicated polarity, an output voltage U_0 with the same polarity will emerge. If U_I is reversed, U_0 will get reversed as well. One could consider U_I to be composed of voltages U_2 and U_1 both with respect to ground: $U_I = U_2 - U_1$. The output will react quite the same as shown above: $U_0 = V \cdot U_I = V \cdot (U_2 - U_1)$.

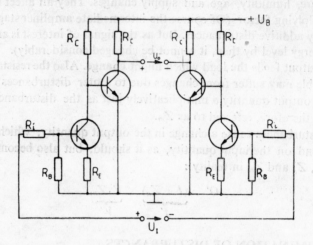

Fig. 10.3.1-1. Symmetrical Difference Amplifier.

In this way operational amplifiers are usually used. They provide two inputs: The input quantity U_2 is processed directly, the other input quantity U_1 is processed inversely.

The amplifier of Fig. 10.3.1-1 as well as all other operational amplifiers employ directly coupled stages. So they may amplify DCs as well as ACs up to high frequencies. This advantage is countered by the disadvantage that the sensitivity towards drifts increases with the number of stages. In order to keep drift effects within tolerable limits, manufacturers undertook enormous efforts to provide differential amplifiers, with stages extremely equal to each other. (However, pure drift free amplification is possible only with pure AC amplifiers, which the various chopper types make use of. But for them, the frequency of the input signal is limited to about 1/10 of the chopper frequency).

Common Mode Rejection

Disturbing voltage U_Z may interfere with the input signals U_2 and U_1 to be amplified (Fig. 10.3.1-2). A disturbing outer strayfield may easily superimpose a disturbing voltage U_Z of the same magnitude and the same sign to both inputs. U_Z is equally picked up because both inputs are usually fed through the same input cable, which contains both input leads very close to each other. So input 2 will consider $U_2 + U_Z$ and input 1 takes $U_1 + U_Z$ into account, resulting in an output voltage:

$$U_0 = [(U_2 + U_Z) - (U_1 + U_Z)] \cdot V_0$$

$$= V_0 (U_2 - U_1)$$

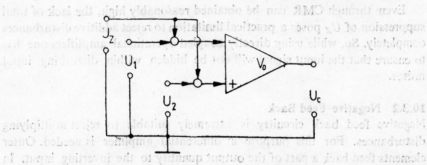

Fig. 10.3.1-2. Rejection of Common Mode Disturbances.

The disturbing voltage U_Z is effectively rejected if the gain for input 1 is the same as for 2. But in fact a slight unsymmetry will remain in reality for the two amplifiers. Usually this is signified by the so-called common mode rejection (CMR) or the common mode rejection ratio (CMRR):

$$\text{CMR} = 20 \lg \frac{V_0}{V_Z} = 20 \lg (\text{CMRR}) \quad \text{with} \quad \text{CMRR} = \frac{V_0}{V_Z}.$$

V_0 is the gain effecting the signal voltage $(U_2 - U_1)$ and V_Z is the gain for the common mode disturbing voltage U_Z' which should be ideally zero. For a short-circuited U_Z the signal gain V_0 can be determined. Afterwards making $U_1 = U_2$ (let both equal zero, for instance) and applying a certain U_Z the gain V_Z may be found. V_0 and V_Z allow to calculate CMR and CMRR.

Ex. The following figures are typical ones as they could be found in a data sheet, e.g., for the operational amplifier µA 741 for general purpose applications.

$$V_0 = \left| \frac{U_0}{U_2 - U_1} \right| \quad \text{e.g., } V_0 = 100000$$

Now let $U_1 = U_2 = 0$

$$V_Z = \left| \frac{U_0}{U_Z} \right| \quad \text{e.g., } V_Z = 10$$

$$\text{CMR} = 20 \lg \frac{V_0}{V_Z} = 20 \lg \frac{100000}{10}$$

$$= 20 \lg 10000 = 80 \text{ dB}$$

$$\text{CMRR} = \frac{V_0}{V_Z} = 10000$$

Because of the arbitrary sign of U_Z, even AC disturbances like hum voltages are effectively rejected. The differential amplifier employs this principle, to cope with common mode disturbances effectively. They are amplified some 10000-times less than the signal of interest.

Even through **CMR** can be obtained reasonably high, the lack of total suppression of U_Z poses a practical limitation to reject additive disturbances completely. So, while using directly coupled operational amplifiers one has to ensure that the input signal will not be hidden within disturbing input noises.

10.3.2   Negative Feed Back

Negative feed back circuitry is extremely suitable to reject multiplying disturbances. For this purpose a differential amplifier is needed. Outer elements feed back a part of the output quantity to the inverting input. In this way only the difference of $(U_2 - U_1)$ is amplified, and the amplifier output produces the voltage U_0 (Fig. 10.3.2-1). The outer feed back network

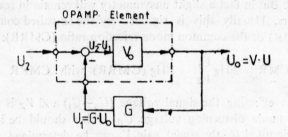

Fig. 10.3.2-1.   Negative Feed Back Circuit.

(G) determines the part U_1 of U_0, which is fed back: $U_1 = G U_0$. As U_1 reduces U_2 the total gain V of the circuit will be reduced compared to V_0 of the actual amplifier element, which is named the open loop gain. V may be obtained in the following way as $V = U_0/U_2$.

$$U_0 = (U_2 - U_1)\, V_0$$

$$\frac{U_0}{V_0} = U_2 - G U_0 \qquad G U_0 = U_1$$

$$U_0\left(\frac{1}{V_0} + G\right) = U_2$$

$$\frac{U_0}{U_2} = \frac{1}{\dfrac{1}{V_0} + G} = \frac{V_0}{1 + G V_0} = V$$

This is the common definition of the gain of the fed back amplifier. In case V_0 is large enough ($V_0 \to \infty$) the gain V' will only depend on the feature G of the outer feed back network:

$$V' = \lim_{V_0 \to \infty} V = \frac{1}{G}$$

This is an extremely notable result, for manufacturers provide components with an open loop gain of $V_0 \geqslant 100000$, even for low cost operational amplifiers.

If V_0 is large enough it may even suffer from changes due to multiplying disturbances. As V_0 is not contained anymore the overall features (i.e. the total gain V) cannot be effected. This is a highly appreciated advantage of feed back techniques. The gain error ε_V' becomes zero for instance, if V_0 is sufficiently large:

$$\varepsilon_V \approx \frac{V'-V}{V'} = G\left(\frac{1}{G} - \frac{1}{\frac{1}{V_0} + G}\right) = 1 - \frac{G}{\frac{1}{V_0} + G}$$

$$\varepsilon_V' = \lim_{V_0 \to \infty} \varepsilon_V = 0$$

As V_0 can easily be designed for a very high value, multiplying disturbances can be rejected almost completely. The only limitations for the use of amplifiers are additive influences, especially for very low input quantities.

10.4   FUNDAMENTAL AMPLIFIER TYPES

Operational amplifiers are elements with a very high (open loop) gain V_0. Usually a part of their output quantity is fed back to the input. In this way the features of the whole circuit can be designed for any convenience.

10.4.1   The Four Basic Circuits

In case the amplifier produces an output voltage U_0, a negative voltage feed back to the input is provided. If the input quantity is a voltage U_i, the feed back voltage U_f is connected in series to the input voltage. This arrangement is called the series voltage feed back circuit. An input voltage U_i produces an input voltage U_0 [Fig. 10.4.1-1 (a)]. U_0 directly follows U_i.

But the circuit may be chosen to produce an output current i_0. For the case when an input voltage U_i should be measured the current part to be fed back needs to be transformed into a voltage first, which is effected with the help of of R_f. The voltage drop U_f is connected in series to the input voltage U_i. We have the so-called series current feed back circuit [Fig. 10.4.1-1 (b)]. i_0 directly follows U_i.

In case the input quantity is a current i_i, and the output quantity a voltage U_0, the feed back quantity needs to be a current i_f. It is obtained from the output voltage with the help of the resistor R_2. By parallel connection i_f is united with i_i. The arrangement is referred to as shunt voltage feed back circuit [Fig. 10.4.1-1 (c)]. U_0 is inverted with respect to i_i.

The last possibility presents the shunt current feed back circuit [Fig. 10.4.1-1 (d)]. A part i_f of the output current i_0 is fed back to the input current i_i, employing a parallel joint. The output current i_0 is inverted with respect to the input current i_i.

Circuits (a) and (b) provide a very high input impedance. So they allow to measure voltages without loading them. Circuits (c) and (d) show a very low input impedance. So they are used as ideal current measurement devices.

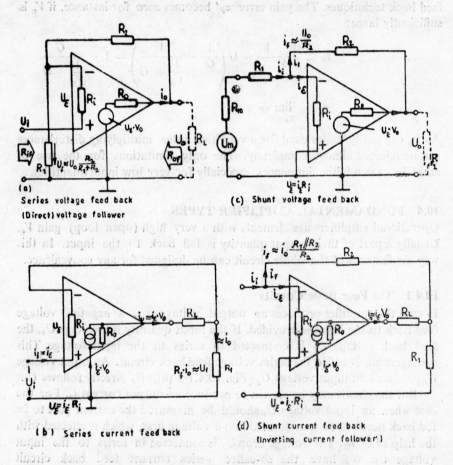

(a)
Series voltage feed back
(Direct) voltage follower

(c) Shunt voltage feed back

(b) Series current feed back

(d) Shunt current feed back
(Inverting current follower)

Fig. 10.4.1-1.   Basic Operational Amplifier Circuits.

The load R_L to be driven, determines whether the output should supply a constant voltage or a constant current. For further voltage processing, circuit (a) or (c) may be selected. Both are constant voltage sources which provide a very low output impedance. But if an ammeter should indicate the measurand, the circuits (b) or (d) may be used. Their outputs are constant current sources with nearly an infinite output impedance.

10.4.2 Example of Calculation for the Voltage Amplifier

The following calculations quantify the abovementioned features. The voltage amplifier, circuit (a) serves as an example.

Take the following assumptions as valid

$$(R_1 + R_2) \gg R_0 \text{ and } R_I \gg R_1$$

The feed back resistors R_1 and R_2 may usually be chosen in this way. Under these conditions the following calculations may easily be effected.

Gain V

The gain V of the whole circuit (a) is certainly $V = U_0/U_i$. Let us express U_0 and U_i in a way that V becomes dependent on the open loop gain V_0 of the amplifier element and the feed back elements R_1 and R_2.

$$U_\varepsilon = U_I - U_f \qquad (1)$$

$$U_f = \frac{R_1}{R_1 + R_2} U_0 \qquad (2) \quad \frac{R_1}{R_1 + R_2} = G \text{ (Feed Back Ratio)}$$

$$U_0 = U_\varepsilon V_0 \qquad (3) \text{ using (1)}$$

$$= (U_I - U_f) V_0 \qquad \text{using (2)}$$

$$= (U_I - G U_0) V_0$$

$$V_0 U_I = U_0 (1 + G V_0)$$

$$\frac{U_0}{U_i} = \frac{V_0}{1 + G V_0} = V = \frac{1}{\dfrac{1}{V_0} + G} \approx \frac{1}{G} = \frac{R_1 + R_2}{R_1}$$

The feed back ratio G is a factor between 0 and 1. G determines which part U_f of U_0 is fed back to the input. As V_0 is usually very high, the gain V is exclusively determined by the outer feed back components R_1 and R_2.

Input Resistance

The total input resistance R_{if} of the series voltage feed back circuit Fig. 10.4.1-1 (a), is obtained from the voltage ratio:

$$\frac{R_{if}}{R_I} = \frac{U_I}{U_\varepsilon}$$

R_I is the input resistance of the actual amplifier element. The voltage drop U_ε across R_I causes the output voltage U_0. But because of the feed back circuit, an input voltage U_I is needed to produce the same U_0. From (1) we obtain

$$U_I = U_\varepsilon + U_f \qquad \text{using (2)}$$

$$= U_\epsilon + \frac{R_1}{R_1 + R_2} U_0 \quad \text{using (3)}$$

$$= U_\epsilon + GV_0 U_\epsilon$$

$$= U_\epsilon (1 + GV_0) \qquad (4)$$

So U_I needs to be $(1 + GV_0)$ times bigger than U_ϵ to produce the same U_0 as U_ϵ does. The relation of U_I/U_ϵ allows to calculate the total input resistances R_{If}.

$$\frac{U_I}{U_\epsilon} = 1 + GV_0 = \frac{R_{If}}{R_I}$$

$$R_{if} = R_i (1 + GV_0)$$

The input resistance R_{if} of the series voltage feed back circuit is considerably increased compared to the internal resistance R_i of the amplifier element.

Ex.   For typical values of $R_I = 1000\,\Omega$, $V_0 = 100000$ and a chosen circuit gain of $V = 10$, $(G = 1/10)$, R_{if} becomes

$$R_{if} = 1000\,\Omega \left(1 + \frac{1}{10}\,100000 \right) \approx 10\ \text{M}\Omega$$

Output Resistance
The output resistance R_{of} of a generator is usually obtained from the measurement results for the unloaded generator and the short-circuited generator. Here it is quite the same. The unloaded amplifier produces the output voltage $U_{0(UL)}$ and the short-circuited output carries the current $i_{0(SC)}$. These quantities are determinants of the internal output resistance R_{of} of the amplifier:

$$R_{of} = \frac{U_{0(UL)}}{i_{0(SC)}} \qquad (5)$$

For the unloaded output the voltage $U_{0(UL)}$ may be determined from equation (3) which is based on the unloaded case.

$$U_{0(UL)} = U_\epsilon \cdot V_0 \qquad (6) \quad \text{together with (4)}$$

$$= \frac{U_I}{1 + GV_0} V_0 \quad (7)$$

The short-circuit current $i_{0(SC)}$ is determined by the amplifier output resistance R_0 and the supply voltage $U_{0(UL)}$ which delivers the current:

$$i_{0(SC)} = \frac{U_{0(UL)}}{R_0} \qquad \text{together with (6)}$$

$$= \frac{U_\epsilon V_0}{R_0}$$

If the output is short-circuited, no feed back takes place ($U_f = 0$). So, according to (1) $U_\varepsilon = U_t$, making

$$i_{0(SC)} = \frac{U_t V_0}{R_0} \qquad (8)$$

Putting (7) and (8) into (5) delivers R_{of} as

$$R_{of} = \frac{U_t}{1 + GV_0} V_0 \frac{R_0}{U_t V_0}$$

$$R_{of} = \frac{R_0}{1 + GV_0}$$

The output resistance R_{of} of the series voltage feed back circuit is considerably reduced compared to the output resistance R_0 of the amplifier element. **Ex.** With a typical R_0 of $100\,\Omega$ and for the same circuit elements as for R_i the total output resistance becomes very low as

$$R_{of} = \frac{100\Omega}{1 + \dfrac{1}{10} 100000} \approx 10\,\text{m}\Omega$$

Linearity
For the ideal amplifier the output voltage U_0 should be directly proportional to the input voltage U_ε. But for the real element it is not (Fig. 10.4.2-1).

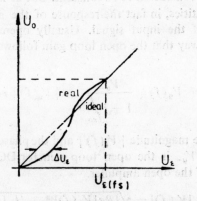

Fig. 10.4.2-1. Linearity Deviation of a Real Amplifier Element.

For the plain amplifier element a linearity error ε_0 may be defined as

$$\varepsilon_0 = \frac{\Delta U_\varepsilon}{U_{\varepsilon(fs)}} \qquad (9)$$

$U_{\varepsilon(fs)}$ is the input voltage which covers the full range of the amplifier element (still avoiding saturation of the output).

If we employ the feed back circuitry, the input voltage $U_{i(fs)}$ is applied instead of $U_{\varepsilon(fs)}$. So the linearity error of the feed back circuit should be defined as

$$\varepsilon_f = \frac{\Delta U_\varepsilon}{U_{i(fs)}}$$

Equation (4) gives $U_{i(fs)}$ which is introduced into ε_f:

$$\varepsilon_f = \frac{\Delta U_\varepsilon}{U_{\varepsilon(fs)}(1 + GV_0)} \text{ with (9) } \varepsilon_f \text{ is obtained as}$$

$$\varepsilon_f = \frac{\varepsilon_0}{1 + GV_0}.$$

The linearity error ε_f of the feed back circuit is $(1 + GV_0)$ times better than the one ϵ_0 of the amplifier element only.
Ex. For $\epsilon_0 = 10\%$, the total linearity becomes extremely good. With the same amplifier data as chosen for R_{lf}, the linearity error becomes

$$\epsilon_f = \frac{0.1}{1 + \dfrac{1}{10}\,100000} \approx 10^{-5}.$$

Bandwidth

So far the gain was defined as ratio of output voltage over input voltage. The quantities were considered to be constant. But as many applications deal with changing quantities, in fact the response of the amplifier depends on the frequency f of the input signal. Usually operational amplifiers are designed in such a way that the open loop gain follows an equation of the form

$$V_0(f) = \frac{V_{0-}}{1 + j\dfrac{f}{f_{co}}} = |V_0(f)|\, e^{j\varphi(f)}$$

$V_0(f)$ represents the magnitude $|V_0(f)|$ and the phase $\varphi(f)$ of the gain at a certain frequency. V_{0-} is the open loop gain for DC quantities, f_{co} is the corner frequency of the open amplifier.

$$|V_0(f)| = \sqrt{\{Re\,[V_0(f)]\}^2 + \{Im\,[V_0(f)]\}^2}$$

$$= \frac{V_{0-}}{\sqrt{1 + (f/f_{co})^2}}$$

$$\tan \varphi = \frac{Im\,[V_0(f)]}{Re\,[V_0(f)]}$$

$$\varphi = \text{arc tan} \frac{-f}{f_{co}}$$

The depiction of $| V_0(f) |$ through BODE-plot makes use of logarithmic scales for the gain axis and the frequency axis, as shown in Fig. 10.4.2-2.

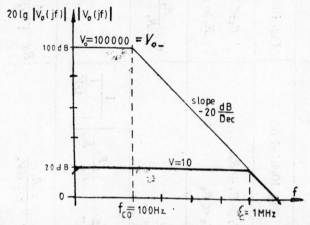

Fig. 10.4.2-2.　Shifting the Critical Frequency by Means of Feed Back.

For low frequencies the gain $| V_0(f) |$ becomes V_{0-}. The frequency response is given as a horizontal asymptote. For high frequencies the gain follows asymptotically to the line $| V_0(f) | = V_{0-} \cdot [f_{co}/f]$. It decreases by 20 dB for ten times increase in frequency ("Roll Off"). The slope of the line equals -20 dB/Decade.

Ex.　The negative feed back takes the gain down from V_0 of the amplifier element to V. At the same time the critical frequency is considerably taken up from f_{co} to f_c, (Fig. 10.4.2-2):

$$ lg \ \frac{f_c}{Hz} - lg \ \frac{f_{co}}{Hz} = lg \ V_0 - lg \ V $$

$$ f_c = \frac{V_0}{V} \ _{co} = (1 + GV_0) f_{co}. $$

The bandwidth is widened to far higher frequencies.

This approach is somewhat simplified. For actual circuit design one should refer to the data sheet of the operational amplifier. Usually the "roll off" increases the slope continuously for increasing frequencies.

10.4.3　Comprehensive Table of Amplifier Features

The previous derivations could be repeated for all feed back circuits (a) to (d) of Fig. 10.4.1-1 in the same way. The results are shown in the following table. Whatever output or input resistance might be necessary for a certain application, the appropriate circuit may be chosen from the table.

It is not substantial whether the output of the operation amplifier is chosen to be a voltage or a current source. However, the right choice may ease the derivation of the results.

Table of static features of operational amplifiers in different negative feed back circuits

Kind of Feed Back	Series Voltage Feed Back	Series Current Feed Back	Shunt Voltage Feed Back	Shunt Current Feed Back
Depiction	Fig. 10.41.-1 (a)	Fig. 10.4.1-1 (b)	Fig. 10.4.1-1 (c)	Fig. 10.4.1-1 (d)
Pre-suppositions	$R_1+R_2 \gg R_0$ $R_i \gg R_1$	$R_f \ll R_0$ $R_f \ll R_i$	$R_i \gg R_1+R_m$ $R_0 \ll R_2$	$R_1 \ll R_0$ $R_2 \gg R_i$
Feed Back Ratio	$G=\dfrac{R}{R_1+R_2}$	$G=\dfrac{R_f}{R_f+R_L+R_0}$	$G=\dfrac{R_i}{R_2}$	$G=\dfrac{R_1}{R_1+R_2}$
Gain	$V=\dfrac{U_0}{U_i}=\dfrac{V_0}{1+GV_0}$	$K=\dfrac{i_0}{U_i}=+\dfrac{1}{R_f}\left[\dfrac{1}{1+\dfrac{1}{GV_0}}\right]$	$V=\dfrac{U_0}{U_m}=-\dfrac{R_2}{R_1+R_m}\left[\dfrac{1}{1+\dfrac{1}{GV_0}}\right]$	$\dfrac{i_0}{i_i}=\dfrac{-V_0}{1+GV_0}$
Input Resistance	$R_{if}=R_i(1+GV_0)$	$R_{if}=R_i(1+GV_0)$	$R_{if}=R_1+\dfrac{R_i}{1+GV_0}$	$R_{if}=\dfrac{R_i}{1+GV_0}$
Output Resistance	$R_{0f}=\dfrac{R_0}{1+GV_0}$	$R_{0f}=R_0(1+GV_0)$	$R_{0f}=\dfrac{R_0}{1+GV_0}$	$R_{0f}=R_0(1+GV_0)$

For all circuits the linearity error is reduced and the bandwidth increased through feedback. Generally,

$$\epsilon_f = \frac{\epsilon_0}{1 + GV_0} \quad \text{and} \quad f_c = (1 + GV_0) f_{co}$$

10.4.4 Actual Amplifier Circuits for Practical Use

Fig. 10.4.4–1 shows a collection of approved measurement amplifiers and their transfer equations specifying them. A few remarks should facilitate the use of these circuits.

Ex. *Approved Analog Measurement Amplifiers*

1: The reference voltage of a zener-diode could be transformed into another voltage of the same stability.

2 to 5: The circuits sensing the bridge currents, ensure a voltage drop across the measurement terminals which is nearly zero. The one (3) sensing the bridge voltage provides no loading.

6: The charge of a capacitor may be indicated with the help of an ordinary panel meter. The amplifier poses nearly no load for the capacitor. The discharge current is less than 10^{-13} A.

7: A chain of biased rectifiers provides a parabola of high precision (0.1%). The output voltage depends on the squared input voltages. For details see Fig. 10.4.4-2 function generator.

8: Input voltages across the inverting and non-inverting inputs cause an output voltage coming up as their algebraic sum and difference. The input resistors allow to weigh each input voltage suitably.

9: This circuit is extremely free from noise and provides a considerably high input impedance. Its common mode rejection is very high.

10: An impedance Z can be converted into its negative value and may even be weighted (n). It may either be voltage or current driven.

11: I_ϵ is negligible compared to the input current I_E. So I_E causes directly the output voltage U_A, coming from a low impedance source.

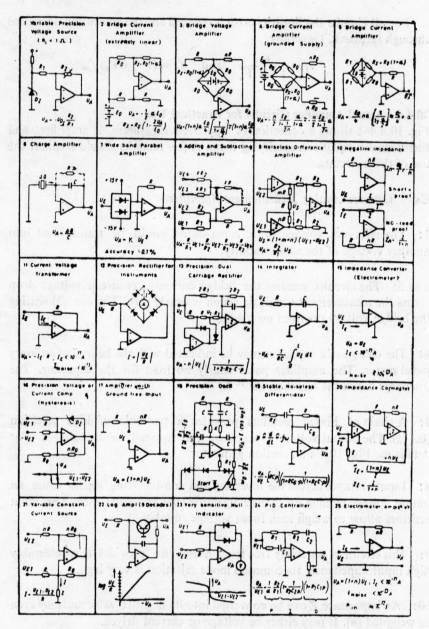

Fig. 10.4.1-1.   Approved Analog Measurement Amplifiers. (After the German
Periodical ELEKTRONIK, Working Sheet No. 6)

12:   The current *I* passing through the instrument is the absolute amount
of the input current. A moving coil instrument will indicate the magnitude
of the mean value.

13:   The output provides the magnitude of the mean of the input also for non sinusoidal voltages. Amplifier 1 allows only positive voltages to pass. To the resulting U_1 half of U_E is added. The sum is integrated.

14:   The integration is better with less amplifier input current, as compared to the one through C.

15:   Impedance converter of electrometer quality ($I_\varepsilon \ll 10^{-11}$ A).

16:   The high open loop gain of operational amplifier allows a precise comparison between voltages or currents. The zener diode prevents saturation to facilitate fast response. The two input voltages need to be of opposite sign. If they are equal for their absolute values (difference less than 1mV), this causes an output step. By feeding back a part of U into the positive input, a hysteresis behaviour may be obtained.

17:   The circuit may be used as a preamplifier for a not grounded transducer voltage U_E. Action principle like circuit 3.

18:   The oscillator provides an exact frequency proportional to $1/RC$. The diodes provide stable output amplitude.

19:   The differentiator tends to amplify noise disturbances. The capacitor C_G prevents saturation of the amplifier for high disturbing input frequencies by integrating them.

20:   The impedance Converter allows to set the input impedance Z_E to certain values within a wide range.

21:   For a changing load Z the supply current I can be maintained at a constant value which is determined by two voltages.

22:   The logarithmic transistor feature within the feed back loop of the amplifier, provides an output voltage proportional to the logarithm of the input voltage.

23:   The zero crossing of $(U_{E1} - U_{E2})$ is sensed. The two diodes prevent amplifier saturation and ensure a fast response.

24:   This circuit combines linear amplification behaviour (P) with the one of integrators (I) and differentiators (D). The P, I and D parts can be set to certain values each.

25:   The circuit allows a gain greater than one. The other features are
the same as of circuit 15.

The reference for these circuits is "Working Sheet No. 6" of the German
periodical ELEKTRONIK.

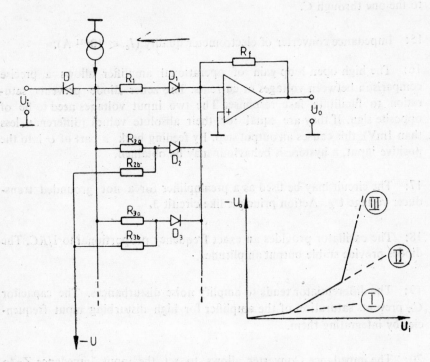

Fig. 10.4.4-2.   Function Generator.

Function Generator

The output voltage U_0 of a simple feed back circuit employing an operational amplifier element is known as

$$U_0 = - \frac{R_f}{R_i} U_i$$

R_f is the feed back resistor and R_i the input resistor. The gain R_f/R_i can
be influenced, for example, by changing R_i with the help of the input
voltage U_i. This is done in Fig. 10.4.4-2 which presents a function generator
providing an output voltage U_0 proportional to U_i^2.

The diode D serves to reject temperature effects. It is fed by a constant
current source which keeps it always in its conducting state via the internal
resistance of the source for U_i. The diode D_1 is kept in this way right at
the brink of becoming conductive. This remains valid for changing surrounding temperatures also. As U_i starts to rise from zero, diode D_1 immediately

starts to conduct allowing an input current through R_1. The output voltage U_0 for the range (I) is determined by the gain R_f/R_1:

$$U_0 = -\frac{R_f}{R_1}\,U_i \quad \text{(I)}$$

The diodes D_2, D_3, etc. are negatively biased, so they are blocked. But if U_i rises high enough, D_2 becomes conductive switching R_{2a} parallel to R_1. In this way the gain of range (II) is increased causing a faster change of U_0:

$$U_0 = -\frac{R_f}{R_1//R_{2a}}\,U_I \quad \text{(II)}$$

Further increasing U_i switches D_3 into the ON-state and in this way R_{3a} gets shunted to the previous input resistors. The slope of $U_0(U_I)$ becomes steeper in range (III) compared to (II):

$$U_0 = -\frac{R_f}{(R_1//R_{2a})//R_{3a}}\,U_I \quad \text{(III)}$$

One can easily imagine that for a sufficient number of branches a parabola $U_0 = a \cdot U_i^2$ may be approached. For reversed diodes and a reversed current source, U_i may be negative even. In combination with two additional inverters all four quadrants can be covered. For this network lines I, II, III of increasing slopes are obtainable. If the diode branches are employed within the feed back loop of R_f, negative slopes are obtainable. Employing both means, allows to approach any curve also up and down going ones for instance a sinusoidal function. In case a very slowly rising U_i is applied the sine is generated having a very slow frequency. This method is used in function generators.

Parabola Multiplier
The multiplication of two variables a and b can be effected employing parabola function generators within the circuit of Fig. 10.4.4-3.

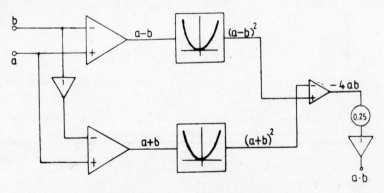

Fig. 10.4.4-3. Parabola Multiplier.

This structure follows directly from the following equations:

$$a \cdot b = -0.25 \, [(a-b)^2 - (a+b)^2]$$
$$= -0.25 \, [a^2 - 2ab + b^2 - a^2 - 2ab - b^2]$$
$$= -0.25 \, [-4ab]$$

The last line proves the validity of the first line. So the structure of this circuit (which realizes this first line directly) certainly allows to obtain the product $a \cdot b$ from the output.

11 Digital Measuring Methods, A/D-Converters

Digital measuring instruments provide basically the same accuracy as the analog ones, because the measurand is normally picked up with the help of a transducer, which is commonly an analog device. However, the resolution of the measurement result can be improved effectively, resulting in extremely low reading uncertainties.

Most physical quantities are subject to steady changes (depending on other variables) characterizing them as analog ones. An analog transducer detects them and converts them into an electrical output quantity, usually a voltage. This is processed further on to achieve a digital indication, as shown in Fig. 11-1. Basically digital measuring instruments are digital voltmeters. Only the transducer determines the kind of measurand to be picked up. An amplifier, being part of the transducer, may step up the voltage $U_X = cX$ to a useful level. The digitalization of U_X is effected by

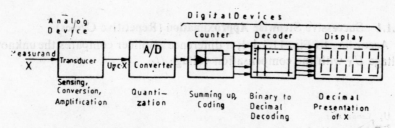

Fig. 11-1. Principle of Digital Measuring Instruments.

consecutive procedures such as quantization, counting and coding. How are these words defined? To quantify means to divide the total voltage range U_{max} into a certain number of steps, providing each the same step-width of ΔU_S which is the resolution of the measured voltage U_X. The accuracy of the sensor limits the resolution. A reasonable maximum number N_{max} of quantization steps should not exceed $N_{max} = U_{max}/\Delta U_S$. These steps are well defined and the instrument counts their number N which is contained within the measurand U_X. The analog-to-digital (A/D) converter provides this number of N pulses within a certain time (serial conversion). The following counter adds them up in this way effecting the actual measurement. But the counting is based on the binary number system. This is due to the fact that it is technically easy to distinguish two different states from each other, for instance the ON and the OFF state of an electronic switch. A binary number is to be coded with respect to a decimal character which one is usually accustomed to. The coding is effected by employing a binary counter. The final counter state represents the measurand as a combination of ON and OFF states of its outputs (parallel code). They may provide a voltage (ON) to the decoder (or not: OFF) which converts the binary information into a decimal one, which is finally displayed. This conversion needs no time. The transfer of data from one number system into another can be effected without loss of accuracy, also its processing. This is the actual advantage of digital instruments over analog ones.

A/D converters employ different methods providing different features.

11.1 COMPENSATION METHODS

The compensation methods basically make use of the balancing principle as it was described for the analog compensator, see chapter 7.1.1, by changing a voltage U_C by exactly known portions. In this way the unknown voltage U_X is approximated until the difference ΔU of both voltages becomes zero. The approximation is effected stepwise either successively within repetitive cycles, or it is gradually done in one go.

The fastest approximation is effected alternatingly (not described here). U_C over- and undershoots U_X alternatingly at each consecutive step. The stepwidth of the next pulse is reduced by a factor of two compared to the previous one, until the deviation of U_C from U_X is sufficiently low.

11.1.1 Successive Stepwise Approximation (Repetitive Cycles)

According to Fig. 11.1.1-1, a difference amplifier compares the unknown voltage U_X with the compensation voltage U_C.

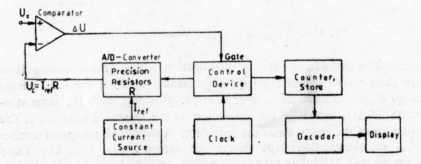

Fig. 11.1.1-1.   Block Diagram of a Digital Voltmeter using Successive Stepwise Approximation Method.

U_C is obtained from a network of precision resistors R which are fed with a constant reference current I_{ref}. After having applied U_x, the comparator result $\Delta U = [U_x - U_C]$ is far from zero. So ΔU opens the gate which controls the clock pulses and allows them to pass to the counter and to the precision resistors. As long as ΔU is different from zero, the clock pulses are used to switch the precision resistors R to higher and higher values, in this way increasing the comparison voltage U_C in small quantities. Once $\Delta U = 0$ the control gate is blocked. The counter output provides a binary number representing the sum of all pulses which were required to balance the instrument. The decoder transfers the binary information to the decimal characters, which are displayed.

Usually a new compensation cycle is started once the balance has been reached

$$(\Delta U = 0 \nearrow U_x = U_C).$$

For this purpose the control device switches the precision resistors back to zero and the balancing procedure is repeated. Meanwhile the previously obtained counterstate is maintained and indicated using a storage memory component. After the next cycle the new measurement result will be indicated. Even for an unchanged measurand the indication suffers from short interruptions of presentation after each cycle.

Disturbances effecting the measurand may easily produce a wrong compensation result. But an input filter can improve this situation considerably. The resolution ΔU of the comparator and the stability of the compensation voltage U_C determine the accuracy of the instrument directly.

11.1.2 Gradual Approximation (Non Repetitive)

The previous method of a digital voltmeter used a clock generator of a constant frequency. But the gradual approximation method of Fig. 11.1.2-1 employs a voltage controlled oscillator (VCO). The frequency of its pulses is

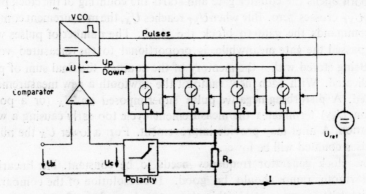

Fig. 11.1.2-1. Simplified Block Diagram of a Digital Voltmeter using Gradual Approximation Method.

proportional to the applied difference voltage $\Delta U = [U_X - U_C]$ of the voltage to be measured (U_X) and the compensation voltage (U_C), which is achieved from a standard resistor R_S. If the compensation is far from balance, the pulse frequency is high. It is low near balance and eventually zero for the balance point. This behaviour features the following procedure: If, for instance, U_C is smaller than U_X, a positive difference voltage ΔU switches the binary coded decimal counters 1 to 4 into their upward position. All incoming pulses switch more and more resistors of the stages 1 to 4 into action this way increasing the current I which passes through the standard resistor R_S. This increases U_C. Thus ΔU decreases and so does the pulse frequency. Near balance, only a very few pulses per unit time arrive, until the last one switches the counter into its final position. This counterstate contains the information as to how many pulses were needed to make U_C equal to U_X. Their number is a measure for U_X. and is indicated. One balancing procedure of this kind is performed and a stable display appears. Only another U_X can cause the instrument to compensate again. But fundamentally there is no difference between this compensator and the one of the previous chapter 11.1.1.

11.2   SINGLE SLOPE METHODS

There are two single slope methods, both employing an integrator. For the first instrument, which effects a voltage to time conversion, a comparison takes place heading for zero balance. It is a compensation device. But the voltage to frequency converter, integrates the measurand U_X in this way gaining a considerable noise attenuation, chapter 11.5.

11.2.1   Voltage to Time Conversion

The single slope method employs a simple A/D conversion. It compares the measurand U_X with a sawtooth voltage U_{ST}, (Fig. 11.2.1-1). The sawtooth opens the counter gate and starts the counting of the clock pulses, once U_{ST} crosses zero. But when U_{ST} reaches U_X, the measurement comparator commands the gate to block the clock. The number of pulses which have passed the gate meanwhile, is proportional to the measured voltage U_X. Being stored within the memory of the counter, the final sum of pulses is indicated. With each new ramp of the sawtooth a new measurement is effected. A disturbing negative pulse superimposed on U_X (or a positive one on U_{ST}) terminates the measurement cycle too early causing a wrong indication. A filter may provide improvement. For a lower U_X the number of pulses counted will be lower.

The clock generator frequency needs to be constant. The linearity of the sawtooth ramp should be good. The resolution of the comparators (and its drift) compensates within the counted number of pulses, because the time instants of start and stop are effected both in the same way.

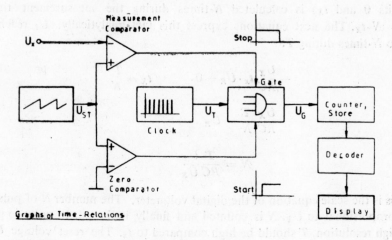

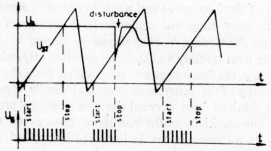

Fig. 11.2.1-1. Digital Voltmeter using Single Slope Method
(Voltage to Time Conversion).

11.2.2 Voltage to Frequency Conversion

The voltage U_x to be measured is converted into a proportional frequency of the integrator output voltage U_I, see Fig. 11.2.2-1:

$$U_I = \frac{1}{RC} \int_0^{t_x} (-U_x)\, dt + U_R$$

$$= \left[-\frac{U_x}{RC} t \right]_0^{t_x} + U_R$$

This voltage U_I follows a straight line commencing with the reset voltage U_R which is applied to the integration capacitor C at each zero crossing of U_I, and the integration starts again. At the same time the zero detector generates the pulses U_C to be counted. The frequency $1/t_x$ of this pulse voltage is proportional to U_x which can be seen from the last equation. A timer opens the gate during the fixed measurement time T and passes on all incoming pulses U_C to the counter. The number N of pulses is high for a high U_x. For a certain U_x the value of the integral (within the time limits 0 and t_X) is calculated N-times during the measurement time $T = N \cdot t_X$. The next equations express this fact analytically. U_I reaches zero N-times during T:

$$-\frac{U_x}{RC} t_x \cdot U_R + 0 \qquad t_x = \frac{T}{N}$$

$$\frac{U_x}{RC} \cdot \frac{T}{N} = U_R$$

$$N = \frac{T}{RC} \frac{U_x}{U_R}$$

This is the scale equation of the digital voltmeter. The number N of pulses is proportional to U_X. N is counted and finally indicated. In order to get a high resolution, T should be high compared to t_x. The reset voltage U_R is actually the reference of the instrument and needs to be well defined and stable. R and C are the outer integrator components. A change of their time constant RC changes the reading N. They should be chosen carefully with respect to their long term stability and their temperature dependence. For high measurands U_x the fast integrator slope suffers from changes, tending to spoil the linearity of the instrument. The periodically repeated gate time T is normally obtained from a crystal timer in order to provide the necessary stability. The sensitivity of the instrument towards all these outer influences limits its accuracy. It can be improved by employing the modified integration methods described in the next chapters. Anyhow one important advantage was realized here: It has comparatively low sensitivity for humm disturbances due to the integration of the measurand. In chapter 11.5, the effect of noise attenuation is shown.

Constant Error voltages U_E, effecting the integrator input, may cause considerable indication errors ϵ_{SS} of the single slope device. If U_E increases U_x, the integration time t_x is unfortunately decreased. In reality this value is smaller than the ideal one: t_x (real) $< t_x$ (ideal).

The values of t_X (ideal) and t_X (real) can be calculated from the integrals for the ideal voltage U_x and the real one ($U_x + U_E$) as

$$t_X \text{(ideal)} = \frac{U_R}{U_x} RC \quad \text{and} \quad t_X \text{(real)} = \frac{U_R}{U_x + U_E} RC$$

Both are functions of U_X. Introducing them into ϵ_{SS} the indication error is obtained also as a function of U_X:

$$\epsilon_{SS} = \left| \frac{t_X \text{(real)} - t_X \text{(ideal)}}{t_X \text{(ideal)}} \right| = \begin{cases} \left| \dfrac{U_E}{U_x - U_E} \right| = \dfrac{U_E/U_x}{1 - U_E/U_x} & \text{for } \dfrac{U_E}{U_x} < 0 \\[3ex] \left| \dfrac{U_E}{U_x + U_E} \right| = \dfrac{U_E/U_x}{1 + U_E/U_x} & \text{for } \dfrac{U_E}{U_x} > 0 \end{cases}$$

The error is depicted, as it depends on U_E/U_X, in Fig. 11.2.2-1: $\epsilon_{SS}\,(U_E/U_X)$. This graph shows that the ratio U_E/U_X should be low to avoid high errors. ϵ_{SS} is high for low measurands U_X. The double integration method allows a much higher U_E in this way expanding the measurement range to far lower measurement voltages of U_X, see chapter 11.4.

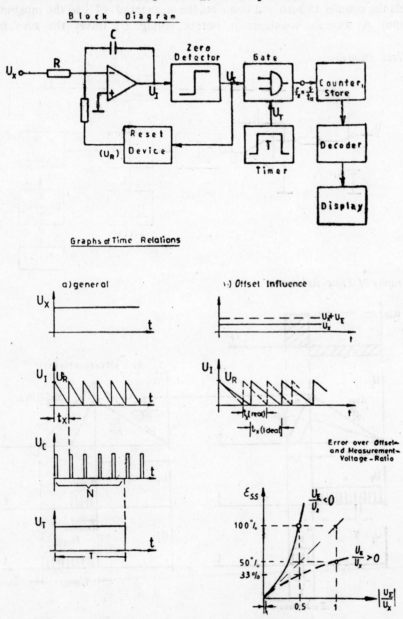

Fig. 11.2.2-1. Digital Voltmeter using Single Slope Method (Voltage to Frequency Conversion).

11.3 DUAL SLOPE METHOD (*U-t*-Conversion)

The main advantage of the dual slope method over the single slope is the high stability of the instrument over long terms, because R and C are not effecting the result, as illustrated here.

At the beginning of the measurement the control device (Fig. 11.3-1), sets the counter to zero and connects the measurand $-U_X$ to the integrator input. As soon as the integrator output voltage U_I leaves the zero line

Block Diagram

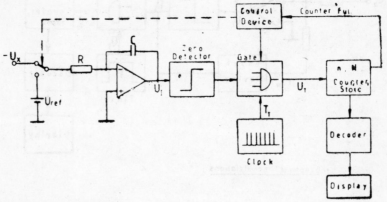

Graphs of Time-Relations

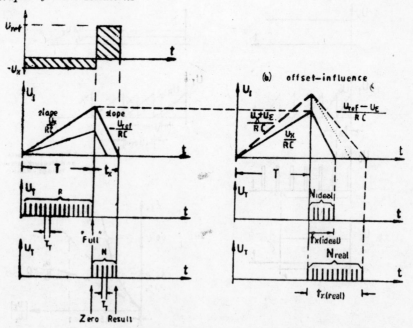

Fig. 11.3-1.   Digital Voltmeter using Dual Slope Method (Voltage to Time Conversion).

(having the slope of U_X/RC) the zero detector opens the control gate and allows to pass the pulses coming from the clock. They are counted by the counter until it reaches its "full" state. This happens independently from the amount of U_X. The counting time T is constant. It is determined by the maximum number n of pulses which the counter is capable to count and the time T_T (period duration of the pulse voltage generated by the clock): $T = n \cdot T_T$. For a counter with four decimals n equals 10^4. The counter "full" state causes a signal which commands the control device to switch the integrator input over to a constant reference voltage U_{ref} which is of opposite sign to U_X. As U_X was negative the positive U_{ref} causes the integrator voltage U_I to decline to zero with a constant slope of $-U_{ref}/RC$

Until this time all incoming pulses U_T are counted as the gate is still open. This new counting starts from zero, for the counter was switched over from its "full" to its "zero" state by the first pulse which arrived after the slope had been changed. Once U_I reaches zero again, the detector blocks the gate and the resulting number N of pulses stored inside the counter is proportional to the measurand U_X. N is decoded and indicated. A new measurement cycle can start. During the next cycle time $(T + t_X)$ the obtained counter state is stored and the indication maintained until a new measurement result is available.

The following derivation proves the relation between the measurand U_X and the indicated number N. According to the action principle (described above) the integrator voltage U_I slopes positively at the beginning (T) and then it slopes negatively (t_X) finally reaching zero after the time of conversion $(T + t_X)$ has passed:

$$U_I (t = T + t_X) = \frac{1}{RC} \int_0^T (+U_X) \, dt + \frac{1}{RC} \int_0^{t_X} (-U_{ref}) \, dt = 0$$

$R \cdot C$ effects the rising ramp as well as the falling, and can be cancelled as can be seen from the last equation. This way one gets rid of RC-changes. This is the reason of the long term stability of this method.

$$t_X = \frac{1}{U_{ref}} \int_0^T U_X \, dt = \frac{T}{U_{ref}} \cdot U_X$$

During the time t_X the number of N pulses are counted: $t_X = N T_T$. Replacing t_X and introducing $T = n T_T$ allows to solve the previous equation for the indicated number N:

$$N = \frac{n T_T}{T_T} \frac{U_X}{U_{ref}} = n \frac{U_X}{U_{ref}}$$

Indeed the indicated figure N is proportional to the measurand U_X. This scale equation shows also that the clock time T_T needs to be the same only during the short time of $(T + t_X)$. Long term changes of T_T cannot effect the result N. This is another feature of the dual slope method.

Concerning an error voltage U_E, it effects the measurement result even more than the same U_E could do for the previously described single slope method. This becomes clear from Fig. 11.3-1 (b). If no U_E was present the ideal counter time $t_{X\,(\text{ideal})}$ would be measurable. But usually an error voltage U_E is present. Let us assume that U_E increases U_X. This causes a a steeper positive slope. Switching over from $-U_X$ to $+U_{\text{ref}}$ cannot reverse the voltage U_E, for it is entirely a feature of the amplifier input. So instead of going down with the dotted slope of $-U_{\text{ref}}/RC$ the integrator slopes down with

$$- (U_{\text{ref}} - U_E)/RC = (-U_{\text{ref}} + U_E)/RC$$

which is slower. This causes a $t_{X\,(\text{real})} > t_{X\,(\text{ideal})}$. So the indication N is wrong. This is the actual set back of the dual slope method. One could define an error ε_{DS}, as it was done for the single slope method (ε_{SS}). ε_{DS} is greater than ε_{SS} if $U_E/U_X > 0$. But in spite of this fact the dual slope method is the most frequently used one. If precautions are taken to ensure a low U_E of the input stage this method is certainly the best one, so far.

11.4 DOUBLE INTEGRATION METHOD (U-f-Conversion)

Latest developments employ a double integration procedure which even allows to tackle the error voltage problem of the input. This advantage is drawn from the peculiarity that ideally the upgoing slope of the integrator output voltage is the same as the downgoing one for constant measurand U_X. Independently from the amount of U_X this is realized by switching over the polarity of U_X once the integrator output voltage has reached a defined level. Unlike the dual slope method this new technique uses the changeable U_X for sloping down instead of a fixed value (which was U_{ref} previously), and instead of sensing zero voltage the comparison is made towards a positive and a negative comparator level U_C.

11.4.1 Action Principle of a Double Integration Voltmeter

The action principle could be described using Fig. 11.4.1-1. U_X is applied to a bridge-switch which allows which to reverse U_X. After the switch the voltage U_{XS} is available for amplification. The resulting voltage U_0 is integrated. The integrator output provides U_I which rises until the positive comparator voltage $+U_C$ is reached (Trigger 4). At this instant of time a pulse triggers the flip flop (T$-$FF 9) which causes to reverse the polarity of U_X. Now U_I starts to decline. Once the negative comparator voltage $-U_C$ is reached, another pulse (now from trigger 6) resets the flip flop T$-$FF 9 to its previous state. One measurement cycle is finished and a new one starts. The blocks 1 to 10 effect the A/D-conversion. The pulses coming

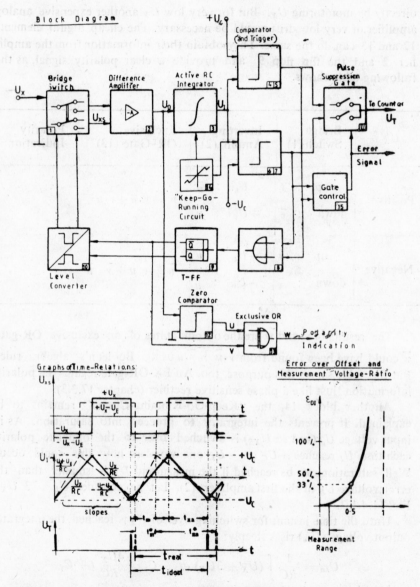

Fig. 11.4.1-1. A/D-Converter of a Digital Voltmeter using Double Integration Method (Voltage to Frequency Conversion).

consecutively from the upper (4; 5) and the lower triggers (6; 7) are used for further processing. The pulses-suppression gate determines whether the pulses come alternatingly from the triggers. If so, every other pulse selected from the gate control 15 is passed to the frequency counter (U_T).

But there are other elements in the circuit not mentioned yet. Blocks 12 and 13 allow to detect the polarity of U_X. Of course one could do this directly by monitoring U_X. But for very low U_X another expensive analog amplifier of very low drift would be necessary. The cheap digital elements 12 and 13 can do the same. They obtain their information from the amplifier 2 and the flip flop 9 and provide a clear polarity signal, as the following table shows.

U_X	Bridge Switch (1)		Inverting Amplif. (2)		Exclusive OR–Gate (13)	Polarity Indication
Positive	up :	v	$-U_0$:	$\bar{u}$	$w = v \cdot \bar{u} + \bar{v} \cdot u$	$+$
	down :	$\bar{v}$	$+U_0$:	u		
Negative	up :	v	$+U_0$:	u	$\bar{w} = v \cdot u + \bar{v} \cdot \bar{u}$	$-$
	down :	$\bar{v}$	$-U_0$:	$\bar{u}$		

The results of w and $\bar{w}$ are the output states of an exclusive OR-gate. $\bar{w}$ could have been found from $\overline{v \cdot \bar{u} + \bar{v} \cdot u}$ using Boolean's algebra rules. But the table serves this purpose, too. An EX-OR-gate regains the polarity information just like a phase sensitive rectifier (chapter 17.3.3).

Another block 14, the "Keep-Go-Running-Circuit" remains to be explained. It prevents the integrator to proceed into saturation. As its input voltage U_0 (and so U_{XS}) is switched over to the opposite polarity each time U_I reaches $+U_C$ or $-U_C$ one wonders how this could occur. Well, saturation can be reached if the measurand U_X is smaller than the error voltage U_E of the first amplifier (2). Let us consider $U_E = 3 \, U_{XS}$, Fig. 11.4.1-2.

Until the time instant for switching over of U_X is reached, the integrator output voltage (U_{Ia}) rises steeply:

$$U_{Ia} = \frac{1}{RC} \int (U_X + 3U_X) \, dt + C_1 = \frac{4U_X}{RC} t + C_1$$

Reaching $+U_C$ the reversing command for U_X is given. However, the integrator output voltage (U_{Ib}) still rises, but with a lower slope (because U_E cannot be reversed):

$$U_{Ib} = \frac{1}{RC} \int (-U_X + 3U_X) \, dt + C_2 = \frac{2U_X}{RC} t + C_2$$

So the saturation would finally be reached and the whole circuit gets out of action. For the saturated integrator a high negative U_X would aggravate this fault further. To avoid this the Keep-Go-Running-Circuit" of block

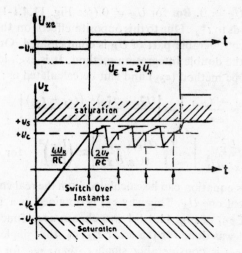

Fig. 11.4.1-2. Effect of Offset Voltage U_E for a Small Measurand U_X.

14 is employed. It senses U_I for voltages

$$+U_C < U_I < +U_S \text{ or } -U_C > U_I > -U_S,$$

in order to monitor whether positive $(+U_S)$ or negative saturation $(-U_S)$ is approached, beyond the positive $(+U_C)$ or the negative $(-U_C)$ comparator voltage. If this is the case, circuit 14 feeds an opposite voltage into the integrator input and forces U_I back to the permitted region between $+U_C$ and $-U_C$. Block 14 gets out of action once U_I has reached a certain lower value which is clearly smaller than U_C. In case the input situation of U_X has not changed meanwhile, the whole procedure of preventing saturation is repeated. This is recognized by the fact that all generated pulses come from the comparator 4 (or 6 respectively) only, but not alternatingly from both comparators. As the indication error could be then considerably high, these pulses are suppressed with the help of block 11 and hence not available for further processing. An error signal indicates that the instrument is out of range. U_X is too small compared to the error voltage U_E.

The time constant RC is contained in both integrator ramps. Due to the same reasons, as shown for the dual slope method, $\tau = RC$ can be cancelled. In this way the long term stability of the instrument is realized without the need of stable components R and C. This method of double integration further provides the feature of being less sensitive towards error influences U_E, such as offsets. This is due to the procedure of processing only every other pulse (U_T). The ideal pulse width t_{ideal} equals the sum of

$$(t_{X1} + t_{X2}) \text{ with } t_{X1} = t_{X2}.$$

This is true for $U_E = 0$. But for $U_E \neq 0$ (see Fig. 11.4.1-1) t_{X1} shortens to t_{X1}^* and t_{X_2} extends to $t_{X_2}^*$. Due to this opposite effects on the up- and down going ramps, a considerable part of U_E is compensated. Only a small error ε_{DI} remains for the double integration method. It is even less than the one for the single slope method (ε_{SS}) and can be calculated as:

$$E_{DI} = \left| \frac{t_{\text{real}} - t_{\text{deal}}}{t_{\text{deal}}} \right| = \left| \frac{(t_{X1}^* + t_{X2}^*) - (t_{X1} + t_{X2})}{t_{X1} + t_{X2}} \right|$$

$$= \left| \frac{U_E^2}{U_X^2 - U_E^2} \right| = \left| \frac{(U_E/U_X)^2}{1 - (U_E/U_X)^2} \right| \approx \left(\frac{U_E}{U_X} \right)^2 \quad \text{for } U_X \gg U_E.$$

The times of this equation can be calculated from the real voltage $(U_X + U_E)$ and from the ideal one U_X. This way ε_{DI} is obtained as a function of U_X. The depiction of ε_{DI} of Fig. 11.4.1-1 was chosen as a function of U_E/U_X. The same way it was done for the error ε_{SS} of the single slope method, (Fig. 11.2.2-1). ε_{DI} is considerably smaller than ε_{SS} for the same ratio U_E/U_X. Or in other words: If the same error for both methods is permitted ($\varepsilon_{DI} = \varepsilon_{SS}$) far smaller voltages U_X are measurable employing this last method than for the single slope method. The measurement range is extended essentially to smaller values of U_X.

11.4.2   Approved Voltage to Frequency Converter

Ex.   An approved circuit makes use of the double integration method. It should be discussed in some details. They are useful for a ny set up of a measurement circuit.

The voltage to frequency converter of Fig. 11.4.2-1 was designed for measurement of voltages ranging from 10 μV to 100 mV without any internal change of scaling. The measurement error does not exceed 5% within the total voltage range for surrounding temperatures of $-30°$ C to $+70°$ C ($20°$ C $\pm 50°$ C). For a limited temperature range, covering $0°$ C to $+40°$C the relative error is less than 2%.

The preamplifier OP 1 provides a very low error voltage U_E. It is the sum of constant offset terms and temperature dependent ones:

$$U_E = \underbrace{U_{OS} + I_{OS} \cdot R}_{\text{const.}} + \underbrace{\frac{du_{OS}}{dT} \Delta T}_{< 05 \mu V/°C} + \underbrace{\frac{di_{OS}}{dT} \Delta T \cdot R}_{\substack{< 10p\, A/°C \\ \downarrow \\ \mu\, 725A\, C}}$$

The constant terms are compensated by employing potentiometer P_1. The drifts of offset voltage and offset current can be kept low if the temperature drift ΔT is kept low. This is effected by mounting OP1, with its TO-5 enclosure, inside a miniature thermostat box which is heated up with the help of an NTC-resistor. The temperature change inside the little chamber is reduced to the 10th part of the outside change. The input resistor R should be of

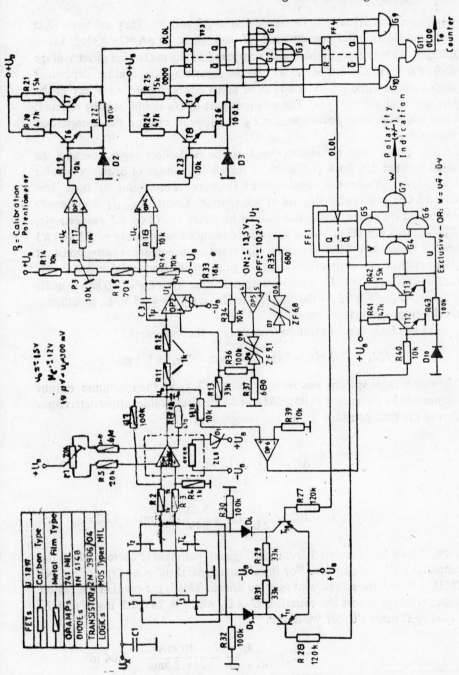

Fig. 11.4.2-1. Realized Circuit of a Voltage to Frequency Converter using Double Integration Method.

low value. R is actually to be considered as $R2$ and $R3$. They are both $1K\Omega$ and produce an offset current drift of less than $10 \text{ pA/°C} \times 5°C \times 1 \text{ k}\Omega = 0.05 \text{ μV}$. This is a negligible value compared to the maximum offset voltage drift of $0.5 \text{ μV/°C} \times 5°C = 2.5 \text{ μV}$. So the remaining temperature dependent error voltage $U_E \approx 2.5 \text{ μV}$ needs to be taken into account, in respect of the measurement voltage U_X. The constant part of U_E could also be rejected, using the calibration potentiometer P_3. It allows to change the comparison voltage U_C.

The switch-over transistors need to be field-effect types, because the small voltage U_X does not allow a threshold voltage of ordinary bipolar transistors. N-channel depletion-FETs were chosen due to their low channel ON-resistances $R_{DS, ON}$ of less than 20 Ohms. $R_{DS, ON}$ is extremely dependent on temperature. But as it is in series to $R2$ or $R3$ respectively, $R_{DS, ON}$ does not effect the total series resistance considerably as $R2$ and $R3$ are 1 k Ohm each. They are metal film resistors which are nearly constant with respect to temperature. The gain of the preamplifier is $A = R7/(R2 + R_{DS, ON}) = 98$. Its highest output voltage (for $U_X = 100\text{mV}$) equals $U_0 = 98 \cdot U_X = 9.8$ V. As the circuit is fed from $U_B = \pm 15$ V, U_0 is sufficiently far from saturation voltage.

Let us choose the maximum frequency of the integrator:

$$f_{max} = 200 \text{ Hz} = 1/(t_{X1} + t_{X2}) \approx 1/2t_{X1} = 1/5 \text{ ms}.$$

The comparator voltage was set to $U_C = \pm 12$ V. So the maximum output change of the integrator voltage $\Delta U_I = 24$ V. These values allow determination of the time constant RC of the integration:

$$\Delta U_I = \frac{1}{RC} \int_0^{t_{X1}} U_0 \, dt$$

$$RC = \frac{U_0}{\Delta U_I} t_{X1} = \frac{9.8\text{V}}{24\text{V}} \cdot 2.5 \text{ ms} = 1.02 \text{ ms}.$$

Now C may be calculated from the highest permissible current which the amplifier OP2 can provide. For the chosen general purpose type of μA 741 (MIL), 10 mA output current may be drawn. This is the case for the fastest voltage change across the capacitor $C = C_3$ which is 24V/2.5 ms. From I_C equal to C times dU_C/dt we obtain

$$C = \frac{I_C}{dU_C/dt} = \frac{10 \text{ mA}}{24\text{V}/2.5\text{ms}} = 1.04 \text{ μF}$$

For this value of $C = 1.04$ μF we calculate R from the time constant $R \cdot C = 1.02 \text{ ms} = \tau$ as

$$R = \frac{\tau}{C} = \frac{1.02 \text{ ms}}{1.04 \text{ }\mu\text{F}} = 980 \text{ }\Omega.$$

For the purpose of compensating temperature effects on τ the temperature coefficients of $C = C_3$ and of $R = (R11 + R12)$ were carefully chosen (see also chapter 17.7.2). But of course, C and R are standard types of 1μF and 1k. The resulting difference of the rated maximum frequency may be compensated with help of the calibration potentiometer P_3.

The integrator voltage may change up to $+ U_C$ or $- U_C$. These trigger levels are determined by the resistors $R14$, $P3$, $R15$, and $R16$. They define the reference voltage $U_C = \pm 12$ V. $U_I = \pm 12$ V is detected by the open loop amplifiers $OP3$ and $OP4$ in combination with the triggers ($T6$, $T7$) and ($T8$, $T9$). The transistors speed up the signal change, which is needed for addressing the following digital elements (which are CMOS components). The diodes $D2$ and $D3$ prevent the bases of transistors $T6$ and $T8$ from being negatively biased with respect to the related emitter. This protection needs to be provided. Otherwise the high negative output voltage $(-14$ V) from $OP3$ or 4 could break over from emitter to base and burn out this junction. Flip flop $FF3$, in combination with the gates $G1$, $G2$, and $G3$, passes the short pulses on to the switch over flip flop $FF1$, which stores the polarity information for the switch-over FET-bridge. $FF1$ is supplied by $U_B = + 15$ V. So its outputs Q and $\bar{Q}$ can provide only 0 V or $+ 15$ V. But for switching the FETs ON or OFF, a gate voltage of 0 V or -12 V should be applied. So a level converter is needed. It is assembled employing the transistors $T10$ and $T11$.

In order to avoid damage of the FETs by a voltage U_X, which might exceed 200 mV, additional circuitry is necessary to protect the input. We do not deal with this case here, but a capacitor $C1$ is provided to block random voltage peaks superimposed on U_X.

All this circuitry mentioned so far, constitutes the voltage to frequency converter. The further processing of the generated pulses from $T7$ and $T9$ is effected by $FF4$. It suppresses every other pulse and in combination with the gates $G9$, $G10$ and $G11$ it allows only such pulses to pass to the counter which are consecutively generated from $T7$ and $T9$. As mentioned earlier, this is not necessarily the case. If U_X is smaller then U_E, the pulses come continuously only from one trigger. Here $T7$ was assumed generating OLOL. But only the first L is passed on. The next one is suppressed because it was not generated from U_X. This pulse comes from the "keep go running" circuit, employing $OP5$. It comes into action because the integrator output exceeds $+ 12$ V. The reasons for such behaviour were given in chapter 11.4.1. $R33$ and $R34$ determine the gain of $OP5$ as $A_5 = 10\text{k}/18\text{k} = 1.8$. Once U_I reaches $(6.8 + 0.7)$. 1.8 V $= 13.5$ V the Zener diodes ZF 6.8 break through and accelerate the output of $OP5$ into negative saturation. This way the branch with $D8$ and $D9$ is switched into its ON state, applying a negative voltage to the non-inverting input of the integrator $OP2$. Its output responds immediately and comes down below U_C until $D\,7$ and $D\,8$ switch off. In case the input situation did not change, the upper level of 13.5V is approached again and the whole procedure is repeated. All pulses are generated only from the upper trigger ($T\,7$) then.

Finally the circuit for the polarity check of U_X needs be described. $OP\,6$ is used in its open state to react sensitively on positive or negative deviations of U_0 from zero. A trigger employing $T\,12$ and $T\,13$ speeds up the pulses for processing them with digital gates $G\,4$, $G\,5$, $G\,6$, and $G\,7$ which realize the exclusive OR: $w = u\bar{v} + \bar{u}v$.

Further detailed information would be needed for covering the circuit completely. But it is felt that it would go beyond the scope of this book. Those details are rather of electronic nature than of direct impact to measurement problems. But implicitly they are explained by the circuit itself.

11.5 INTEGRATOR ATTENUATION OF AC DISTURBANCES

Ripple voltages or harmonics originated from phase cut control devices may interfere with voltage measurements. They are usually superimposed on the measurand U_X and tend to cause errors. All these noises are commonly induced from power lines and therefore their frequencies are (50 or 60) Hz and integer multiples of this fundamental frequency (higher harmonics: see also FOURIER-analysis, chapter 14). The following derivation shows that for all measurement methods, employing an integration of the measurand U_X, the noise disturbances are effectively rejected if the gate time (or measurement time) is chosen as $T = 1/f_n$ or integer multiples of that f_n is the frequency of the interferring noise. The single slope method as it was employed for a voltage to frequency conversion (see chapter 11.2.2) may serve as an example for the following calculations.

The frequency f_X as it depends on the measurand U_X is

$$f_X = \frac{1}{RC} \frac{U_X}{U_R} \quad \left(\text{with } f_X = \frac{1}{t_X}, \text{ chapter 11.2.2}\right)$$

The indicated figure N resulted as

$$N = T \cdot f_X.$$

If a ripple noise is superimposed to the measurand, U_X cannot remain constant but becomes a function of time and so does f_X, as shown in Fig. 11.5-1.

From the graph, one can read the time dependent frequency as

$$f_X = f_{X0} + f_r \cos \omega_n \cdot t$$

The instrument effects an integration during the time $(t_2 - t_1)$ and indicates U_X as the number N which is proportional to U_X and so to f_X:

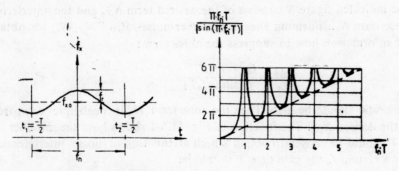

The symbols of this depiction are :

f_{X0} — Desired (or nominal value) of measurement frequency
f_X — Instaneous value of the measurement frequency
f_r — Frequency range
f_n — Noise frequency
T — Gate time.

Fig. 11.5-1. Attenuation of AC–Disturbances using Integration Principles
of Appropriate Measurement Time T.

$$N = \int_{t_1}^{t_2} f_X \, dt$$

The gate time $T = t_2 - t_1$ should be chosen equal to the period duration
$1/f_n$ of the noise voltage. For this case the attenuation of the noise becomes
considerably great, as shown below:

$$N = \int_{-T/2}^{+T/2} (f_{X0} + f_r \cos \omega_n \cdot t) \, dt$$

$$= f_{X0} \cdot T + f_r \int_{-T/2}^{T/2} \cos \omega_n \cdot t \, dt$$

$$= f_{X0} \cdot T + \frac{f_r}{\omega_n} \sin (\omega_n t) \Big|_{-T/2}^{+T/2}$$

$$= f_{X0} \cdot T + \frac{2 f_r}{2 \pi f_n} \sin \left(2 \pi f_n \frac{T}{2} \right)$$

$$= \underbrace{f_{X0} \, T}_{N_{X0}} + f_r \underbrace{\frac{\sin (\pi f_n T)}{\pi f_n}}_{N_n}$$

The indicated figure N consists of the desired term N_{X_0} and the interfering noise term N_n. Forming the "signal-over-noise-ratio" N_{X_0}/N_n, we obtain the information how to suppress the noise term:

$$\frac{N_{X_0}}{N_n} = \frac{f_{X_0}}{f_r} \frac{\pi f_n T}{\sin (\pi f_n T)}$$

This ratio should be high. Then the noise term N_n is negligible compared to the desired N_{X_0}. The function of Fig. 11.5-1 shows how the product of $f_n \cdot T$ should be chosen to obtain a high attenuation of noise interferences. For a certain f_n the gate time T should be

$$T = k/f_n \text{ for } k = 1, 2, 3, \dots$$

In this case the noise attenuation is infinity and only the desired term N_{X_0} is indicated. Concerning this integrator a high value of the k would cite better results than low k. But a higher k extends the measurement time. For low U_X this needs to be tolerated in order to gain sufficient resolution. In case $f_n \cdot T$ is not exactly an integer multiple of unit, this does not matter so much for higher products $f_n \cdot T$ as it does for lower ones.

12 Digital Processing Elements

Digital processing can be effected with a few basic circuits. They are triggers, gates, and multivibrators (flip-flops).

A trigger senses a certain voltage level. If the input voltage rises and reaches finally this trigger level, the output switches over from its 0 state to L state. Going back to smaller input voltages causes the trigger to switch back when another level is reached which is lower than the first one. The difference is called the hysteresis of the trigger.

Gates are quite simple arrangements of a few transistors. With three basic elements realizing the logic interconnections of inversion, AND and OR (see Fig. 12-1) all other gate features of NAND, NOR, EXOR, etc. can be realized employing these fundamental elements.

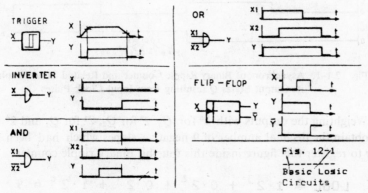

Fig. 12-1. Basic Logic Circuits.

These gates are available with many more inputs. They allow to assemble quite complex logic circuits, but for details, reference to basic digital electronics should be made.

These gates allow even to set up memory features. But ready assembled binary counters are exclusively in use. They allow to perform the most basic measurement task: counting.

12.1 BINARY COUNTER BITS AND CODING

A counter is made up from a series of bits, each having a trigger input T and two outputs Q and $\bar{Q}$. An input change from logic L to O might flip the outputs Q ond $\bar{Q}$ into their states of L and O respectively. Another input change might flop them back to their previous states O and L, see Fig. 12.1-1. Such a flip-flop element may resume these two different states. It is of binary nature and can store two different informations. We may label them as YES and NO, or ON and OFF, or L and O.

The series circuit of four such bits may provide 16 different states of all output combinations Q_0 to Q_3 being available as a parallel information. The outputs Q_3, Q_2 Q_1, and Q_0 may, for instance, provide the states LOOL. This binary "word" indicates that 9 negative slopes had been consecutively applied to the clock input C rippling the counter up to this final state.

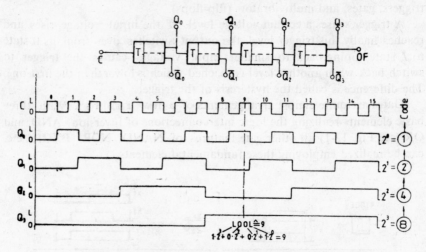

Fig. 12.1-1. Asynchronous Binary Ripple Counter and Related Time Graph of its Output States Q resulting from Input Clock Pulses C.

Weighting the outputs with 2^3 for Q_3, 2^2 for Q_2, 2^1 for Q_1 and 2^0 for Q_0 we obtain the decimal number of 9 negative slopes which had been necessary to register this figure inside this four bit binary ripple counter:

$$\text{LOOL} \cong 1 \cdot 2^3 + 0 \cdot 2^2 + 0 \cdot 2^1 + 1 \cdot 2^0 = 9$$

$$\underbrace{⑧ \qquad ④ \qquad ② \qquad ①}_{\text{Code}}$$

Due to the binary weights the counter codes a decimal number. In this case we say that the 8-4-2-1 code was employed. The process of counting is at the same time closely linked to the binary coding.

The four bit counter allows to distinguish 16 different states of output combinations: it can count from 0 to 15. The 16th negative slope resets the counter to its original state 0000.

Usually we are accustomed to use the decimal system for counting. Due to this fact it is quite reasonable to employ a combination of gates which may trigger the counter back to its zero state after 10 pulses have been counted (0 to 9). A binary counter equipped with such additional logic circuitry is called a binary coded decimal counter, short BCD-counter. The maximum number of its output combinations is 10.

A counter with a capability to count up to 999 pulses consists of three BCD counter components, see Fig. 12.1-2. The input AND gate may enable

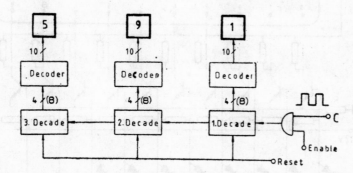

Fig. 12.1-2. 3-Digit Binary Coded Decimal (BCD)-Counter.

the counting procedure for certain time intervals. The incoming clock pulses (at C) are stored into the first decade. The 10th pulse resets it to zero and produces a negative slope which is the first slope counted within the second decade. This procedure may continue until 99 pulses are stored. The 100-th one resets the previous two decades to 0 and gives a first pulse into the third decade, etc. until for example 591 pulses are counted, as depicted here.

12.2 BCD-DECODER AND CHARACTER DISPLAY

Ex. The four outputs of each BCD-counter element carry the stored information which should be indicated. A NIXIE-display provides ten inputs. They allow to address 10 different cathodes which are shaped as 0, 1, 2,...9. These figures are positioned one behind another. They all have a common anode. The whole assembly is housed inside a glass tube which is filled with a gas. If one of the cathodes is powered the gas around it is excited to produce light radiation and a bright appearance of its shape is visible. For the counter of the previous chapter three such NIXIE tubes are employed. The information to be displayed needs to be provided with the help of 10 connection leads for each display unit.

The BCD-counter elements have four Q outputs and four $\bar{Q}$ outputs. For such a configuration a diode array, as shown in Fig. 12.2-1 may be used as a decoder. It acts as a cross bar distribution panel. If for instance the 9 should be displayed the far right transistor is switched ON providing nearly zero potential to cathode 9. As the anode carries a potential of $U_B =$ 200 V the 9 is triggered. How is the transistor addressed? For the 9 the Q output situation is **LOOL** and consequently the one of $\bar{Q}$ is **OLLO** (the

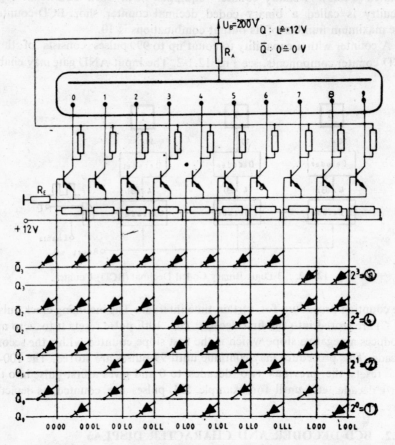

Fig. 12.2-1.   Circuit of a Diode Decoder Matrix and a Display Tube
(NIXIE-type) for Decoding the Stored Information of
a BCD-Counter and Indicating it.

inverted information). Let us assume that L is represented by + 12 V and 0 by OV. All horizontal lines carrying L information are connected to the vertical "9" line via diodes. Their cathodes are on positive 12 volts. That means they are all blocked and so the positive supply voltage of + 12 V biases the base of the transistor positively with respect to its emitter. This causes the collector-emitter-path to become conductive, and ground potential is applied to cathode 9 to make it visible. If only one bit changes at least one diode switches ON pulling the voltage of the vertical "9" line to zero which causes the base of the transistor to be negatively biased. This switches the transistor OFF and the 9 disappears from the display.

Usually a BCD-counter is enhoused together with its decoder inside the same enclosure. Seven segment displays have partly replaced the NIXIE tubes. In this case only seven output lines are needed (plus ground). Of course the decoder is of different internal structure. But the user of these elements need not care for the hardware composition inside the components. He only needs to meet the rated input and output conditions.

12.3 D/A CONVERTERS

To store an information inside a digital memory provides the advantage of being absolutely decisive. But often the contents of a counter are needed for further processing.

Ex. An analog device may, for instance, be powered proportionally to the stored number. In this case the digital information needs to be converted into an analog one. A digital to analog converter usually sums up four currents coming from the Q-outputs of the four counter bits which might provide a voltage ($Q = L = +12$ V) or not ($\bar{Q} = 0 = 0$ V). The binary code word for decimal 9 was LOOL. It demands that

Q_3 is	to be weighted by a factor of 2^3	:	$1 \cdot 8$	$= 8$
Q_2 is not	to be weighted by a factor of 2^2	:	$0 \cdot 4$	$= -$
Q_1 is not	to be weighted by a factor of 2^1	:	$0 \cdot 2$	$= -$
Q_0 is	to be weighted by a factor of 2^0	:	$1 \cdot 1$	$= 1$

$$9$$

Ex. This weighting is effected with the help of resistors $R/8$, $R/4$, $R/2$ and R, see Fig. 12.3-1. They effect the currents which are provided by the counter

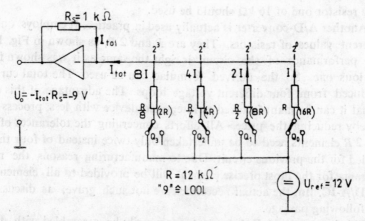

Fig. 12.3-1. D/A Converter with Weighting Resistors of R, $R/2$, $R/4$ and $R/8$.

outputs Q_3 to Q_0. If R was chosen as 12 k ohms and the Q-voltage might be $+12$ V, the current from the Q_0 output will be $I = 1\text{mA}$. For Q_1 it might be $2I = 2\text{ mA}$ due to the weighting resistor of $R/2 = 6$ k ohms, for Q_2 current is $4I = 4\text{mA}$, and for Q_3 current is $8I = 8$ mA. But for the depicted "9" situation the switches Q_3, Q_2, Q_1 and Q_0 are in their positions LOOL and so only $8I = 8$ mA plus $I = 1$ mA will be considered. They make up the total current $I_{tot} = 9$ mA. Due to the feature of an operational amplifier its output produces the same current, passing it through the feedback resistor $R_S = 1\,\text{k}\Omega$. So this circuit produces the output voltage of $U_0 = -I_{tot} \cdot R_S = -9$ V due to the inversion of the o. amp. (The highest possible $U_0 = -15$ V. This happens for LLLL which is the binary word for decimal 15).

The procedure of the previous calculation shows that the total current I_{tot} and the amplifier output voltage U_0 may be generally calculated as

$$I_{tot} = (Q_3 \cdot 8 + Q_2 \cdot 4 + Q_1 \cdot 2 + Q_0 \cdot 1)\, I \quad \text{with} \quad I = \frac{U_{ref}}{R}$$

$$U_0 = -R_S \cdot I_{tot} = -\frac{R_S}{R} \cdot U_{ref} (8Q_3 + 4Q_2 + 2Q_1 + 1Q_0)$$

If the resistors are differently chosen, such as $16\,R$, $8\,R$, $4\,R$ and $2\,R$ the previous R needs to be replaced by $16\,R$ and the output voltage can be calculated from the formula:

$$U_0 = -\frac{R_S}{2R} \cdot U_{ref} (Q_3 + \tfrac{1}{2} Q_2 + \tfrac{1}{4} Q_1 + \tfrac{1}{8} Q_0).$$

There is no particular difference, though the output voltage U_0 equals $-9\text{V}/16$ only for the previous example of decimal "9". If again -9 V should be available R_S needs to be sixteen times higher. That means instead of a $1\,\text{k}\Omega$ resistor one of 16 kΩ should be used.

Another A/D-converter is actually used in practice. It employs only two different values of resistors. They are R and $2\,R$, as shown in Fig. 12.3-2. The performance of the circuit is right the same as it was shown for the previous one. So the derived formulae can be used. The total current is produced from four different voltage loops. The advantage of this circuit is that it can be manufactured as integrated device with less process steps, thereby reducing the price. All efforts concerning the tolerances of the R and $2\,R$ elements need to be undertaken only twice instead of four times as needed for the previous circuit. Due to manufacturing reasons the needed tolerance for the most precise element will be provided to all elements for the D/A-IC. But the actual necessities are not such grave, as discussed in the following passage.

This last type of a D/A converter can easily be assembled with discrete resistor components. The tolerances of the resistors need to be lowest for the most significant bit in order to prevent the least significant bit from being hidden within the tolerances of the more significant bits. For the n-th

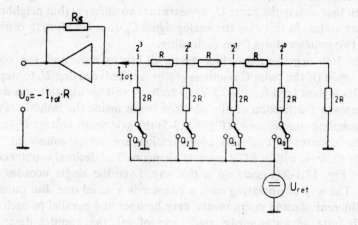

Fig. 12.3-2.   D/A Converter with Weighting Resistors of R and $2R$.

digit the tolerance of the referred resistances needs to be not wider than $\Delta R/R = 1/2^M$. For the fourth digit a tolerance of 6% can be permitted, for the digit 2^{10} it is 0.1%. These are demands which are realistic.

Finally the typical output signal of a D/A converter equipped with four bits can be shown, see Fig. 12.3-3.

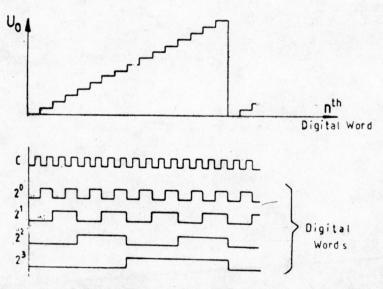

Fig. 12.3-3.   Output States of a 4-Bit Counter and Output Voltage of a Connected D/A Converter.

The clockpulses C are continuously counted rippling up the output voltage U_0 of the attached D/A converter. The steps are evenly spaced if the resistor tolerances are small enough. If they are too rough it might happen that nearly the same U_0 appears for two different (but neighbouring) counter states. In this case the analog signal U_0 does not allow to distinguish these two counter states from each other.

The 16th negative pulse slope resets all bits to zero and the counting starts again (if the pulse C continues to be applied) causing U_0 to step down from its highest level (e.g. $+15$ V) to zero. A voltage like this was used for the stepwise x-deflection of the electron beam inside the cathode ray tube of a sampling oscilloscope, (Fig. 8.2.3-5) the saw tooth voltage U_{ST}.

The counter bits of this configuration are series connected. This is usually the case within BCD counter elements. The decimal counter connection of Fig. 12.1-2 is made up in this way. Even the single decades are in series. The way of operating such a counter is a serial one. But quite often the different decade components may be operated parallel to each other. This is quite advantageously made use of for the counter described in chapter 13.2.

13 Instruments with Digital Data Input

The digital processing elements, as described i n chapter 12, are used to compose measuring instruments of various types. They may cope up with different measurands and different surrounding conditions. The scope of use is manifold. Counters and timers are most frequent applications. They may also be used as components of more complex measurement equipment, such as an electronic ampere-hour meter as it will be presented in chapter 13.3. It is meant for monitoring the available battery charge of an automotive vehicle. This instrument partly makes use of analog processing electronics, too. But also the electronic sensing of the battery temperature will be employed.

Digital action principles provide the advantages of being decisive and not being dependent on drift problems. But mathematical operations require elaborate logic circuits to realize complex calculation procedures. They may take quite some time to obtain a result. These setbacks need not to be accepted entirely by employing analog principles. But they may suffer from limited accuracy. However, digital and analog functional elements are usually combined to facilitate an acceptable compromise which fits the technical and the economic needs. The ampere hour meter, (chapter 13.3), serves as an example.

The designer of measurement equipment should be aware of the basic features of both techniques. They are fundamentally compared in chapter 1.1.3 and referred to in chapters 11 and 12. They facilitate the assessment whether digital or analog approaches should be applied (and in what combination). But the following passages deal only with digital instruments.

13.1 COUNTERS, TIMERS
The fundamental operation of counting events is employed to realize the following instruments either for frequency or time measurements.

13.1.1 Event Counter
Counter units can be used as event counters if the events are presented as pulses. Every time an event occurs, the display changes by one unit. Fig. 13.1.1-1 may be used to present an application.

In a beverage factory, the bottles should be counted which are automatically filled on the production belt over the period of eight hours.

The bottles may interrupt a light beam inside the detector. Once a bottle has passed the previous state is regained. This way a pulse is generated. Another one appears once the next bottle passes the detector. These pulses are usually picked up from a light sensitive diode or a transistor. They need to be amplified to a level which is far above any disturbance which might interfere, in order to obtain a high signal-to-noise ratio. A shaper provides for a fast signal change to which the digital elements may respond i.e. the negative pulse slope. All pulses arriving from the input

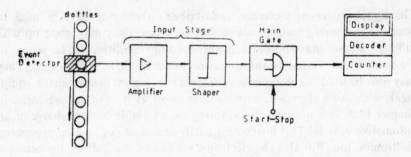

Fig. 13.1.1-1.   Event Counter.

stage are allowed to pass the main gate after the start command was given. They are added up within the counter until the stop command arrives. The display allows to take the reading of the manufactured bottles during the time which has elapsed since the counter was started.

The capacity N_C of the counter is the possible number of output states minus one. A calculation formula can be given for N_C as

$$N_C = B^n - 1$$

Ex. B is the numerical base on which the counter stages operate, i.e. 10 for a decimal counter, n is the number of stages. If a counter with 4 decades is employed, it can count up to $N_{CD} = 10^4 - 1 = 9999$. If a binary counter is used which contains 16 bits, the capacity comes up to $N_{CB} = 2^{16} - 1 = 65535$. (By the way: a BCD counter with 4 decades contains 4 bits per decade, times 4 stages coming up to a total number of 16 bits. The low capacity N_{CD} compared to N_{CB} which was realized with the same number of bits shows the "waste" of bits employing decimal counter stages instead of binary ones. But fortunately IC-technology allows this "waste" because the difference of price for both kinds of components is not high).

13.1.2 Frequency Counter
The events of the previous counter may arrive evenly spaced in time. This means that the pulse voltage which is applied to the event input provides

a constant frequency. If the gate is opened for exactly one second, the number of pulses which pass during this time, presents the frequency of the pulse voltage, for the frequency f is defined as number of events N per second s:

$$f = \frac{N}{1s} = \frac{k \cdot N}{k \cdot s} \qquad [f] = \frac{1}{s} = 1\,Hz$$

If the frequency is low the gate time should be longer than one second in order to obtain sufficient accuracy, see also chapter 13.4. Within k seconds the number of pulses (kN) will be counted. Still the frequency can be

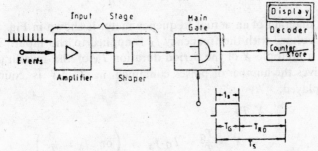

Fig. 13.1.2-1. Action Principle of a Frequency Counter.

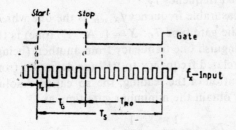

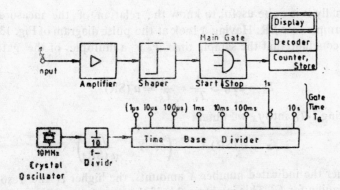

Fig. 13.1.2-2. Block Diagram of a Frequency Meter.

found in the same way as the last equation shows. k may usually be changed in steps of 10. The gate time is chosen high ($k = 1, 10, 100,...$) for low frequencies and it is chosen low ($k = 1/10, 1/100, 1/1000,...$) for higher frequencies to be measured. The instrument (see Fig. 13.1.2-1) is similar to the one of the previous depiction.

The read out time T_{RO} is needed to present the measurement result which was obtained during the gate time T_G. The time in which one sample of the measurand can be taken is the sample time T_S. The rate of samples SR is the inverse sample time. For a low sample rate SR the time for one sample will be long.

$$T_S = T_G + T_{RO} \qquad \mathrm{SR} = \frac{1}{T_S}$$

The block diagram of an actual frequency meter is shown in Fig. 13.1.2-2. The pulse voltage with the frequency f_x is applied to the input. The gate time T_G is a multiple X of the period duration T_x of the measurand. This factor X gives the number of pulses contained in T_G. X is counted and finally displayed.

$$X \cdot T_x = T_G$$

$$X = \frac{T_G}{T_x} = T_G \cdot f_x \qquad \left(\text{or} \quad f_x = \frac{X}{T_G} \right)$$

This scale equation shows that the indication value X is indeed proportional to the measured frequency f_x.

The lowest measurable frequency $f_{x,\,\min}$ is the one which produces only one pulse during the gate time T_G: $X = 1$. A "1" (one) is the least decisive possibility to distinguish one frequency from another (being very close to each other). The related frequency to "1" (as deviation from zero frequency) gives the resolution of the reading, the so called resolution frequency f_{res}. For $X = 1$ we obtain the minimum gate time as

$$T_G = \frac{1}{f_{x\,\min}} = \frac{1}{f_{res}}$$

Occasionally it is quite useful to know the relation of the measurand f_x to the sample rate SR. Having a look at the pulse diagram of Fig. 13.1.2-2, one can conclude that the sample time T_S is a multiple of the gate time T_G

$$T_S = aT_G \Rightarrow \frac{1}{T_G} = \frac{a}{T_S} = a\,(\mathrm{SR})$$

Introducing $1/T_G$ into f_x we obtain

$$f_x = a\,(\mathrm{SR})\,X \qquad \left(\text{or} \quad X = \frac{f_x}{a\,(\mathrm{SR})} \right)$$

The higher the indicated number X amounts, the higher is the resolution of the reading for f_x. This is obtainable if the sample rate is chosen low, which means a long sample time T_S. This way of interpretation asks for a long gate time T_G for gaining a high resolution.

If the meter is designed for human read out, a sample rate of 0.2 to 1 sample per second is feasible. For higher rates another store could be employed which takes over the counter result right after it was obtained. This would allow an immediate new measurement.

The gate time is usually obtained from a stable crystal oscillator of a high frequency. Divider decades step down the frequency for instance to 0.5 Hz (see the depicted connection of the start/stop input of the main gate, Fig. 13.1.2-2). In this case the gate time is $T_G = 1$ s if the pulse time is used for opening the gate and the pause time as read out time $T_{RO} = 1$ s. These internal details usually do not concern the user of the instrument. He gets indication only of the gate time which is exclusively of interest. It can be chosen freely up to 10 s.

13.1.3 Period Duration Meter and Pulse Width Meter

Both instruments employ basically the same action principle of the time measurement. When comparing the previous frequency meter of Fig. 13.1.2-2 with the period duration meter of Fig. 13.1.3-1, it becomes obvious

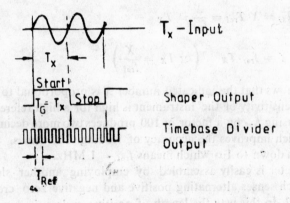

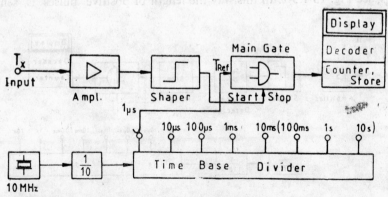

Fig. 13.1.3-1. Block Diagram of a Period Duration Meter.

that only the gate inputs have been interchanged. Previously the start/stop commands were given from a crystal time base. Now they come from the signal to be measured. The period duration measurement method is especially employed when the frequency f_X of the measurand is low. In this case a frequency measurement suffers from a poor accuracy. But the measurement of the period duration $T_X = 1/f_X$ tackles this problem. This is true the other way round, too. So frequency and time measurement instruments supplement each other to obtain acceptable accuracy. This is the reason why both methods are frequently realized within the same instrument.

The period duration T_X of an AC voltage is given by the time which elapses between two identical instantaneous values of successive cycles, i.e. neighbouring zero cross over points of the same slope. A shaper produces the gate singal, see fig. 13.1.3-1, which provides the start command during the period duration T_X and the stop command right after T_X has passed. Within this gate time $T_G = T_X$ the reference pulses are passed over to the counter. The period duration T_{ref} of the short reference pulses is precisely defined. They are derived from a very stable frequency of a quartz crystal oscillator by employing divider decades. The gate time is a multiple X of $T_{ref} = 1/f_{ref}$.

$$T_G = X \, T_{ref} = \frac{X}{f_{ref}} = T_X$$

$$X = f_{ref} \cdot T_X \qquad \left(\text{or } T_X = \frac{X}{f_{ref}} \right)$$

The scale equation shows that the indicated number X is proportional to the measurand T_X. The sensitivity of the instrument is high for a high reference frequency f_{ref}. Increasing f_{ref} by a factor of 100 produces two more decimals at the indication which improves the accuracy of reading 100 times. T_{ref} may be freely chosen down to $1\mu s$ which means $f_{ref} = 1$ MHz.

A pulse width meter is easily assembled by employing another slope detector $(+/-)$ which senses alternating positive and negative zero crossings, see Fig. 13.1.3-2. In this way the length of positive pulses is sensed.

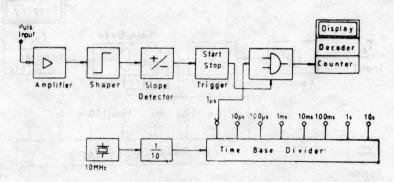

Fig. 13.1.3-2   Block Diagram of a Pulse Width Measuring Instrument.

Changing the slope polarity to $(-/+)$ allows to measure the length of negative pulses. As the switches for selecting the slope polarity are separate components for the two slopes, they both could be set either to $(+/+)$ or to $(-/-)$. In these cases the instrument performs a period duration measurement and is identical to an instrument as it was presented in Fig. 13.1.3-1.

13.1.4 Frequency Ratio Measuring Instrument

An instrument for frequency ratio measurements is not in need of a time base, see Fig. 13.1.4-1. Assume the low frequency signal f_{LOW} is used to start and stop the measurement. During the gate time T_G the number of X pulses of the high frequency signal f_{HIGH} is counted. Assuming a mark to space ratio of 1 : 1 for f_{LOW} the following equations are valid (with $f = 1/T$):

$$T_G = \frac{1}{2} T_{LOW} = X \cdot T_{HIGH}$$

$$X = \frac{1}{2} \frac{f_{HIGH}}{f_{LOW}}$$

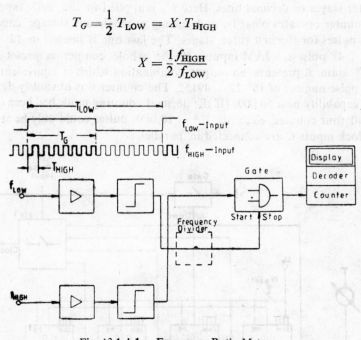

Fig. 13.1.4-1. Frequency Ratio Meter.

Employing a frequency divider within the start/stop connection allows to step down f_{LOW} to a lower value. If this is done for instance with the help of one divider decade the reading X is increased by a factor of 10. In this way the accuracy of the measurement is 10 times higher. One more decimal appears at the display. Of course the decimal point is shifted by one digit to the left in this case.

13.2   ACTUAL 16 BIT BINARY UP/DOWN COUNTER WITH D/A CONVERTER

Ex.   The Up/Down counter of Fig. 13.2-1 is just an event counter. Consequently a time base is missing. It may count incoming pulses upwards or downwards. The presentation of this circuit may be taken for a tutorial in order to illustrate the concepts discussed so far. In addition to the actual counter, a D/A converter is attached to have an analog reading of the counter contents. The following passages describe the function of all blocks in detail. In this way the reader can understand the design of the hardware circuit which employs CMOS components from the 4000-series only. The gates of Fig. 13.2-1 are just blocks. They will be realized using the basic logic elements of NAND and NOR gates. The gate performance can be realized from Fig. 13.2-2 which presents the complete hardware.

Four counter elements are employed. They may be operated as binary counter stages or decimal ones. Here U_B is applied to the B/D input. So the counter operates binarily, each having a maximum storage capability of 16 pulses for the first three stages. The last one is limited to 12 pulses $= (8 + 4)$ pulsess, s JAM inputs. If the whole counter is preset to its "full" state it presents an output combination which is equivalent to a total pulse number of $16^3 \cdot 12 = 49152$. The counter was obviously designed for a capability near 50,000. (If the decimal counting mode had been chosen for all four counter elements $10^4 = 10,000$ pulses could only be stored.) All clock inputs C are connected in parallel.

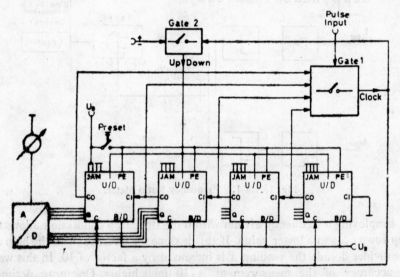

Fig. 13.2-1.   Block Diagram of a Binary UP/DOWN—Pulse-Counter with attached D/A Converter.

But only that counter element which is grounded at its carry input CI responds to a clock pulse. If the first counter is "full" it provides at its carry output CO ground potential for the carry input CI of the next stage, in this way enabling it to accept the next clock pulse at its C-input. The counter cannot overflow. If 49152 pulses are counted gate 1 is blocked, not allowing the counter to switch to its zero state with the next pulse. This counter provides an up or down counting feasibility (U/D) which is commanded by the $(+/-)$ input. But the change command is only allowed to pass gate 2 if no clock pulse is given at the same time. In case this coincidence occurs the available clock pulse is counted still in that direction which was available before the command for polarity change arrived.

Ex. For the indication with the help of a penal meter a resolution (stepwidth) of $15V/49152 = 0.3$ mV is certainly not readable. This is the reason why only the last stages are connected to the D/A-converter. In this way a resolution of $15V/(12 \cdot 16) = 78$ mV is obtained. The stepwidth is 0.52% of full scale deflection. This is certainly sufficient to make the user of the instrument feel that a perfect analog presentation of the measurand is available. The sense behind these measures will be explained in the following chapter 13.3 using the application of an electronic ampere-hour-meter which monitors the available charge of an accumulator. Fig. 13.2-2 shows the actual hardware circuit of the previously described Up/Down counter including the gate. The D/A converter was explained in detail in chapter 12.3.

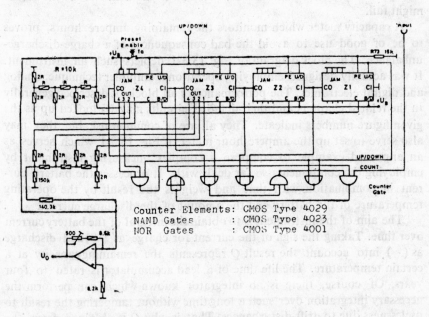

Counter Elements: CMOS Type 4029
NAND Gates   : CMOS Type 4023
NOR Gates   : CMOS Type 4001

Fig. 13.2-2.   Binary Up/Down-Counter and D/A Converter.

13.3 ELECTRONICAL AMPERE HOUR METER FOR BATTERY CHECK OF AUTOMOTIVE VEHICLES

Ex. The tremendous density of traffic in big cities causes an increasing demand for safer vehicles. This challenge is partly met by employing additional instruments which allow to process more data than the driver normally could do. In case of any danger a warning is provided. Electrical power consumers like halogen head lights aim also for more safety. Electronic or electro-mechanical apparatus such as transistor ignition, engine revolutions meter, electronic injection etc. improve the operation reliability of the vehicle. Other consumers aim for means of comfort or entertainment. There are airconditioners, radio receivers, cassette recorders, etc. Whatever kind of performance these instruments may provide, they pose considerable loads to the battery which might discharge beyond control. During the last 25 years the average electrical consumer load in a saloon car has increased from about 150 W to 350 W. But the battery capacity could not be designed to catch up to the same degree and also the dynamo performance was not sufficiently improved. So the power balance is quite often negative. This may cause a severe problem, especially as during rush hours the driving time has dropped to just 50% of the operation time. During the stillstands of the vehicle its engine runs just idle and the recharge current is therefore poor anyhow. So it takes no wonder if the next start of the engine might fail.

A capacity meter which monitors the remaining ampere hours, proves to be of good use to avoid the bad consequences of a charge-discharge-unbalance. The block diagram of Fig. 13.3-1 depicts such an instrument. It was actually designed employing operational amplifier technique, analog and digital electronics. The performance of all blocks is described generally in the related chapters and their detailed circuits may be looked up as the given figure numbers indicate. They all are of common use, but they may also serve to set up the ampere hour meter of Fig. 13.3-1 which serves as an approved example to compose a complex measuring instrument by employing the components so far dealt with. It processes the battery current in combination with time and weights the result by the operating temperature of the battery and the amount of the discharge current.

The aim of the instrument is to obtain the integral of the battery current over time. Taking the sign of the current for charge as $(+)$ and discharge as $(-)$ into account the result Q represents the remaining capacity at a certain temperature. The life time of a lead accumulator is rated to four years. Of course, there is no integrator known which can perform the necessary integration over such a long time without tampering the result to uselessness due to drift disturbances. That is why Q is obtained from the sum of partial integrals. The partial integration time $(t_n - t_{n-1})$ does not exceed that limit beyond which drift influences may be felt considerably.

$$Q = \int\limits_{0}^{4\ years} idt = \int\limits_{0}^{t_2} idt + \int\limits_{t_1}^{t_2} idt + \ldots + \int\limits_{t_n}^{4\ years} idt$$

Of course still some drift disturbance might occur. But if the instrument is included in the regular maintenance check it could be set to indicate the "full" battery state because at these checks a full battery can be provided. The current range to be covered reaches from about 20 mA to 200 A. The low value is valid even for the not running vehicle. The clock still demands the battery. Also self discharge needs to be considered. The high current is needed to start the engine. So a current ratio of 1: 10000 needs to be tackled within one range of the A/D converter. see Fig. 11.4.2-1, being part of Fig. 13.3-1.

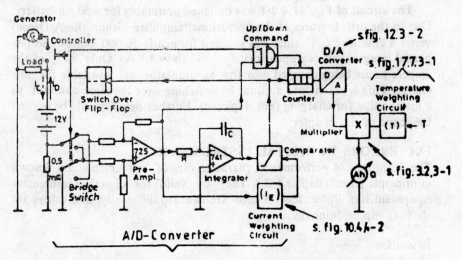

Fig. 13.3-1   Electronic Ampere Hour Meter for Lead Accumulators.

A measurement resistor of 0.5 mΩ is employed to obtain a voltage proportional to the battery current. It ranges from 10 µV to 100 mV. Such voltage drop can be allowed for measurement purposes. It is preamplified by an operational amplifier (725) of extreme low drift and is integrated up to a certain comparator level. Once it is reached, a pulse is generated indicating that one partial integral was processed. This pulse is counted upwards if it was derived from the charge current. If it was obtained from a discharge current the counter will be commanded to count downwards. At the same time the comparator pulse switches over the polarity of the input voltage in this way making use of the double integration method for the A/D-conversion, see chapter 11.4. The instantaneous counter contents represents the accessable charge which should be indicated. For this purpose a D/A-converter is employed. See Fig. 13.2-2 for details. It is known that

the available charge depends on temperature. The coefficient to be considered amounts to about 1% change of the remaining charge per degree
temperature change. As the operating range reaches from −30°C to +70°C,
a temperature measurement needs to be effected, for instance, by employing
the circuit of Fig. 17.7.3-2. The multiplier of Fig. 3.2.3-1 performs the
temperature weighting of the counter contents, in this way the proper
indication Q of the remaining ampere hours is provided.

One detail remains to be mentioned. Due to the nature of the accumulator, the accessible charge Q depends on the amount of the discharge
current. It drops to one third of nominal capacity for the starting current
of 200 A. The function of remaining charge over current is an empirical
one and is realized by employing a diode function generator, see Fig.
10.4.4-2. It shifts the positive and the negative comparator level providing
for the current weighting this way.

The circuit of Fig. 11.4.2-1 was designed preferably for a 50 Ah-battery.
Due to the drift features of the operational amplifier within the A/D-con-
verter a counter "full" state was realized for nearly 50,000 pulses. So one
partial integral values approximately 1 m Ah = 3.6 As. Only 90% of the
ampere hours being charged into the accumulator are available for discharge. This is taken into account by switching over the time constant to
a lower value for charging (not depicted). Further details would be beyond
the scope of this chapter.

13.4 ERRORS IN DIGITAL INSTRUMENTS

The definitions of performance characteristics of instruments are known
in principle from chapter 1.2.5. They are valid for digital measurement
equipment, too. But a few of them are handled differently, and others are
showing up additionally.

Resolution

Ex. The resolution r is that change of the measurand which can cause a
change of one digit, within the least significant unit. A digital voltmeter
with four and a half display units may read for instance 08.73 V in the
200.00 V range (maximum 199.99V). In this case a change to 8.74 V or
8.72 V would cause a change of reading by 1 digit within the least significant decimal place. The resolution is

$$r = \frac{\Delta U}{1_{\text{digit}}} = \pm \frac{0.01 \text{ V}}{1_{\text{digit}}}$$

Usually r is given only as 0.01 V. In the 2000.0 V-range the same measurement voltage would give a reading of 008.7 V. In order to achieve a change
of 1 digit for the least significant decimal place, the voltage needs to be
changed at least to 8.8 V or 8.6 V. The resolution is $r = 0.1$ V then. It
depends obviously on the measurement range, which should be chosen
appropriately for the measurand. Only this way a good absolute resolution
is obtainable.

In the first passage there appeared the technical term of half a digit. This is an additional "1" for the first display unit. It is the most significant bit. Its availability expands the measurement range of the instrument by a factor of two, i.e. 199.99 instead of 99.99. This digit cannot indicate characters other than "1".

Ex. If the instrument is a frequency meter, the resolution will be given as $r = \Delta f / \text{digit}$. Assume $f_x = 1$ MHz should be measured with a resolution of $f_{res} = 0.1$ Hz. What number of decimal displays N_D will be required? The range selector was set to $f_{range} = 1$ MHz.

$$f_X = f_{range} = 10^{n_{range}} \text{ Hz} = 10^6 \text{ Hz} \quad n_{range} = 6$$
$$\Delta f = f_{res} \quad = 10^{n_{res}} \text{ Hz} = 10^{-1} \text{ Hz} \quad n_{res} = -1$$

The required number of displays is

$$N_D = n_{range} - n_{res} \quad N_D = 6 - (-1) = 7$$

Quantization Error

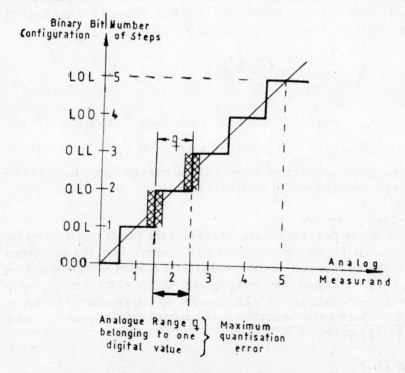

Fig. 13.4-1.   The Quantization Error due to the Related Stepwidth of the Analog Measurand.

The quantization error is expressed in terms of the smallest unit to be counted. It was earlier mentioned in chapter 11. Fig. 13.4-1 depicts the nature of this error. It emerges as consequence of the internal comparison of the measurand with a well defined number of parts of the reference.

These discrete equal steps are counted. At the balance point their sum represents the measurand, employing a binary code. Thus the analog measurand needs to be changed by one quant q to cause the digital output information to change by one bit. A three bit counter was chosen to be the base for these explanations. So the next 3-bit word can be obtained only if the analog quantity changes at least by one quant. The quantization error of the ideal instrument is $E_q = \pm q/2$ or $\pm 1/2$ LSB (of the least significant bit). In practice the decision levels are always somewhat inaccurate. This fact derives mostly from tolerances of different components. Imprecise ranges of q to both sides may cause an indication of 1 or 3 instead of the actual 2. So each reading is uncertain by ± 1 digit. This error cannot be eliminated. It is due to the nature of quantization.

A sensible design of an instrument provides a resolution r of the digital processing elements which is not higher than the one of the analog sensor. The quantization error and the resolution are closely related in this case, and so the question for the necessary number of bits n arises if a certain resolution should be obtained. (Implicitly this problem was handled in chapter 13.1.1, too. N_c is the counter capacity).

$$r \geqslant \frac{1}{N_c} = \frac{1}{B^n - 1}$$

Ex.

$r = 1.00\% = 0.01$ $\quad 2^n \geqslant 101$ $\quad\frown\quad n \geqslant 6.658$ $\quad$ chosen $n = 7$ bits

$r = 0.10\% = 0.001$ $\quad 2^n \geqslant 1001$ $\quad\frown\quad n \geqslant 9.967$ $\quad$ chosen $n = 10$ bits

To perform the operation of quantifying and coding a signal, the A/D converter requires the so called conversion time.

Gate Time Uncertainty

The frequency f_X is determined by counting the sum of pulses during the gate time T_G. The time instant of the "ON" command for the gate comes arbitrarily with respect to the position of the counted negative slopes, see Fig. 13.4-2. In the first case during T_G only "3" pulses are detected, but in the second case there are "4". The direction of this uncertainty cannot be predicted. So, whatever indication is obtained the result includes ± 1 pulse due to the gate time shift.

Total Error

Single errors are only given if they cannot be included into the total error, such as drift errors due to age and temperature or calibration errors. It has become quite a common practice to cite a total error for digital instruments. It is usually given as percentage value ε_{max} which may be calculated as

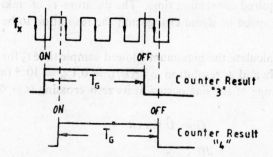

Fig. 13.4-2. Uncertainity of Counter Result due to Gate Time "Shift".

shown below. It contains the actual measurand (U) and the full scale deflection (U_{range}). Both quantities are related to their rated relative errors ε_1 and ε_2 respectively.

Ex. The previously cited voltmeter measured $U = 8.73$ V within the 200 V range. The manufacturer may rate the relative errors of any voltage (U) as $\varepsilon_1 = 0.02\%$ and of full scale deflection (U_{range}) as $\varepsilon_2 = 0.01\%$. So the total maximum error of the digital voltmeter is for this range and this measurand, the absolute value ΔU_{max} from which the total relative error ε_{max} results, as follows:

$$\Delta U_{max} = \pm \, (\Delta U_1 + \Delta U_2)$$
$$= \pm \, (\varepsilon_1 \, U + \varepsilon_2 \, U_{range})$$
$$= \pm \, (2 \times 10^{-4} \times 8.73\text{V} + 1 \times 10^{-4} \times 200 \text{ V})$$
$$= \pm \, (1.746 \text{ mV} \pm 20 \text{ mV}) = \pm \, 21.746 \text{ mV}$$

$$\varepsilon_{max} = \frac{\Delta U_{max}}{U} = \pm \frac{21.746 \text{ mV}}{8.73\text{V}} = 0.25\%$$

This result indicates a considerable high total error ε_{max} compared to the rated errors ε_1 and ε_2 which are usually cited in the data sheet of the instrument. The need arises (as for analog instruments, too) to match the range of instrument near to the actual measurand.

Sample (Aperture) Error and Shannon Theorem
For fast changing measurands A/D-converters should provide very short conversion times. The sampling method avoids this setback and allows to perform measurements even for very high frequencies. This is effected by taking rapid samples of the measurand, see chapter 8.2.3 (sampling

oscilloscopes). An electronic switch is closed for a very short time, the sample or aperture time t_S. During this time the measurand is considered to stay nearly constant. Then the switch may open and the sample is held for the required conversion time. The duration t_S of taking a sample depends on the speed of signal change and the resolution to be realized.

Ex. Let us calculate the maximum allowed sample time t_S for a sinusoidal voltage of 1 kHz and a resolution of $\Delta U/U = 0.1\% = 10^{-3}$ ($n = 10$ bits). The fastest change of the sine occurs at its zero crossing ($t = 0$).

$$U = \hat{U} \sin \omega t$$

$$\frac{dU}{dt} = \hat{U} \omega \cos \omega t$$

$$\left(\frac{\Delta U}{\Delta t}\right)_{max} = \hat{U} \omega \qquad \Delta t = t_S$$

$$t_S = \frac{\Delta U/\hat{U}}{\omega}$$

$$= \frac{10^{-3}}{2\pi \cdot 10^3} \text{ s} = 159 \text{ ns}$$

After the sample time t_S was determined we need to know how often a sample should be taken. Whatever periodical signal we might sample, it is always composed of harmonics, see chapter 14: Harmonic Analysis. The highest harmonic (f_H) to be taken into account determines the sample rate. It is a sinusoidal function. Knowing this, it is sufficient to have two coordinates within its period duration in order to recover this sine completely without loss of information. All the more this is the case for lower harmonics, too. So the whole periodical signal can be regained though only samples of it are taken. These thoughts were developed firstly by NYQUIST and later technically realized by SHANNON. In practice the sample frequency should be taken as $f_S \geqslant 5f_H$ (instead of $2f_H$) due to noise effects and non ideal filter characteristics of the sampling device. If the highest frequency $f_H = 1$ kHz, the sample frequency should be at least $f_S = 5$ kHz. But one can take the first sample at any time instant. The next one needs not to be taken within the same period duration of the highest frequency. (If so, this would need faster response of the measurement equipment than the measurand changes). Instead one can take the next sample after lots of periods have passed. Of course this sample needs to be shifted by a known time interval within this period under view. In this way only a low sample rate is required. Of course, the signal to be sampled should be periodically available, see also Fig. 8.2.3-5.

Generally the transfer functions of the sample switch and the hold circuit allow to calculate the relative sample error as

$$\varepsilon_S = Si\,\frac{\pi f_H}{f_S} - 1 \qquad\qquad f_H = \text{highest frequency within the}$$
$$\text{signal to be sampled.}$$

$$\approx (-)\frac{(\pi f_H/f_S)^2}{3!} + \cdots \qquad\qquad f_S = \text{sample frequency}$$

The second formula gives the first term of the related infinite series which approaches ε_S sufficiently accurate if ε_S is small enough. This term gives a formula for the highest frequency allowed if a certain sample error ε_S is permitted to occur:

$$\frac{f_H}{f_S} = \frac{1}{\pi}\sqrt{\sigma\,|\,\varepsilon_S\,|}$$

Ex. If the sample error ε_S (which describes the permissible deviation from the actual signal) for f_H is assumed as 1 %, f_H should not exceed 7.8 % of f_S. It should be stressed again that ε_S concerns only the distortion of f_H. Lower harmonics of the periodical non-sinusoidal signal are transferred with far higher accuracy. An other interpretation of the formula allows to calculate f_S as 12.8 kHz for $f_H = 1$ kHz if $\varepsilon_S = 1$ %. This f_S is needed if the shape of the sampled curve should be regained within $T_H = 1/f_H = 1$ ms, i.e. for fast (nearly instantaneous) conversion.

13.5 BASIC CALCULATIONS FOR DIGITAL VOLTMETER DESIGN

This chapter is for understanding the basic calculations especially for the time base and the counter of a digital voltmeter. The voltage U_x to be measured is converted into a frequency which is counted. The circuit of Fig. 11.2.2-1 is used, employing the single slope method for A/D conversion. The blocks depicting the timer and the counter of the frequency meter to be connected are shown in Fig. 13.1.2-2.

The voltage to frequency converter may generate the frequency $f_{x\,\text{max}} = 10$ MHz for the maximum input voltage of $U_{x\,\text{max}} = 1000$ V. The digital frequency meter is to be employed as indicator which should provide a resolution of 10^{-6} (equivalent to 10 Hz out of 10 MHz).

Let us determine the number N_D of display units required:

$$f_{x\,\text{max}} = f_{\text{range}} = 10^{n_{\text{range}}}\ \text{Hz} = 10^7\ \text{Hz} \qquad n_{\text{range}} = 7$$
$$f_{\text{res}} = 10^{n_{\text{res}}}\ \text{Hz} = 10^1\ \text{Hz} \qquad n_{\text{res}} = 1$$
$$N_D = n_{\text{range}} - n_{\text{res}}$$
$$= 7 - 1 = 6$$

Now the gate time T_G should be determined. It is the time during which such a number of pulses may pass in order to indicate the lowest frequency

$f_{X \min}$ which is the resolution-frequency f_{res}.

$$T_G = \frac{1}{f_{\text{res}}} = \frac{1}{10\text{MHz} \cdot 10^{-6}} = 0.1 \text{ s}$$

The clock frequency of the crystal oscillator is $f_C = 40$ MHz. What number of divider stages will be needed, and what divider ratio do they need to provide? Let us assume that the gate time T_G is followed by an equal pause time. So the start stop signal is a pulse voltage of a pulse to space ratio of 1/1. This means the time base needs to provide the reference frequency

$$f_{\text{ref}} = \frac{1}{2T_G} = \frac{1}{2 \times 0.1 \text{ s}} = 5 \text{ Hz}$$

So the frequency of 40 MHz needs to be scaled down to 5 Hz defining the total divider ratio as

$$R = \frac{f_C}{f_{\text{ref}}} = \frac{40\text{MHz}}{5\text{Hz}} = \underbrace{8}_{2^{n_b}} \times \underbrace{10^6}_{10^{n_d}} = R$$

The total divider ratio R consists of a binary term and a decimal term. The binary one is realized employing $n_b = 3$ binary divider stages and the decimal one by help of $n_d = 6$ decimal dividers.

The absolute resolution ΔU of the voltage measurement follows from the counter capacity N_c. In order to reach the relative resolution of 10^{-6}, the capacity of the counter needs to be

$$N_c = 10^6 - 1$$

$$\Rightarrow \Delta U = \frac{U_{x \max}}{N_c + 1} = \frac{1000 \text{ V}}{10^6} = 1 \text{ mV}$$

$(N_c + 1)$ is the number of decision levels available.

Assume the input voltage range to be changed from 1000 V to 1 V by changing only the gate time. All display units should be used. What will be the absolute voltage resolution $\Delta U'$?

$$\Delta U' = \frac{\dfrac{1000 \text{ V}}{1000}}{10^6} = 1 \text{ μV}$$

The maximum frequency of the A/D converter will be reduced to

$$f'_{x \max} = \frac{10 \text{ MHz}}{1000} = 10 \text{ kHz} = \underbrace{10^4 \text{ Hz}}_{10^{n \text{ range}}} \Rightarrow n_{\text{range}} = 4$$

The number N_D of display units remains 6 as before. Consequently n_{res} results as

$$n_{\text{res}} = n_{\text{range}} - N_D = 4 - 6 = -2$$

This allows to calculate also the absolute frequency resolution and the relative one as

$$f'_{res} = 10^{n\ res}\ Hz = 10^{-2}\ Hz$$

$$\frac{f'_{res}}{f'_{x\ max}} = \frac{10^{-2}\ Hz}{10^4\ Hz} = 10^{-6}$$

Of course the resolution of 10^{-6} was maintained by changing the gate time from T_G into T'_G. But a 1000 times smaller input range is obtained for U_X. The gate time calculates as

$$T'_G = \frac{1}{f'_{res}} = \frac{1}{10^{-2}\ Hz} = 100\ s$$

(The referred reference frequency is $f_{ref} = 1/2T_G = 5$ mHz).

This gate time is certainly too long. To avoid this the period duration measurement method should be employed which is effected by interchanging the gate inputs of the instrument. The circuit is shown in Fig. 13.1.3-1. If a high clock frequency is chosen the gate time may be considerably shortened.

14 Harmonic Analysis of Periodical Signals (Fourier)

As we know average and effective values characterize non-sinusoidal periodic signals. They can, for instance, be measured by moving coil instruments, and moving iron instruments respectively. At the same time these signals are also characterized by the fact that their energy is distributed over a wide range of frequencies, ranging from the first harmonic (fundamental wave) to harmonics of higher order (overtones). They can be investigated for their amplitudes with the help of instruments employing resonance circuits, which can be tuned with respect to their natural frequency.

In order to sustain the waveshape of such signals on a transmission line, the bandwidth of the line should be wide enough. Otherwise they suffer a change of their shape which means a partial loss of information. This is, of course, of interest for telecommunications. For power transmission, however, higher harmonics cause reactive power, which loads the line additionally due to a higher current with respect to the one which is needed to transmit the active power only.

It is desirable to have a precise description of the energy distribution over frequency of the different signals to be transmitted, irrespective of whether a good transfer of all harmonics is needed or not. This chapter aims at gaining knowledge quantitatively about the amplitude distribution of different non-sinusoidal quantities with respect to the frequency of their harmonics.

Any periodical function of non-sinusoidal shape can be composed of a sum of sine and cosine functions. The single terms provide information about amplitude and phase of the harmonics contained within the spectrum of the function under investigation. There is always a fundamental wave (the first harmonic) and lots of overtones (higher harmonics) the frequencies of which are integer multiples of the fundamental frequency.

These facts highlight the need for sufficient bandwidth of a transfer channel which is used for transmitting non-sinusoidal signals. If the limiting frequency is not high enough, upper harmonics are cut off, producing a distorted output signal.

14.1 PRAGMATIC APPROACH TO FOURIER ANALYSIS

The validity of the Fourier analysis may be proved by sketching the different harmonics and adding their instantaneous values. The resulting curve will be the investigated nonsinusoidal function. The amplitudes of the harmonics can either be analytically obtained, using the Fourier formulae or they are measured using a selective amplifier. Both methods will be discussed. Our approach takes the validity of the Fourier analysis for granted. The more harmonics are considered, the more closely the sum of their instantaneous values meets the actual instantaneous value of the curve under test. This pragmatic proof of Fourier's thoughts convinces and eventually allows the use for practical applications. The graphs of Fig. 14.1-1 present an example. By superimposing more and more harmonics the block curve $I(t)$ may be generated finally.

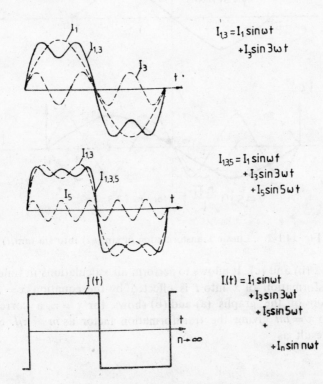

$$I_{1,3} = I_1 \sin \omega t$$
$$+ I_3 \sin 3\omega t$$

$$I_{1,3,5} = I_1 \sin \omega t$$
$$+ I_3 \sin 3\omega t$$
$$+ I_5 \sin 5\omega t$$

$$I(t) = I_1 \sin \omega t$$
$$+ I_3 \sin 3\omega t$$
$$+ I_5 \sin 5\omega t$$
$$+$$
$$+ I_n \sin n\omega t$$

Fig. 14.1-1.   Superimposition of Sinusoidal Currents I_1 to I_n (Reversed Approach to FOURIER Analysis: Synthesis)

Before actually studying the analysis, a useful linear transformation of the variable x should be introduced. The time dependent quantity $y(t)$ needs to be considered within its periodic duration $T = 2l$, see Fig. 14.1-2 (a). The first harmonic (fundamental wave) has the same period. Normally y is used as a function of x: $y = \sin x$, see Fig. 14.1-2 (b). But technical deal mostly with time functions. So a transformation of the 'riable x into the variable of time t proves to be quite useful, see

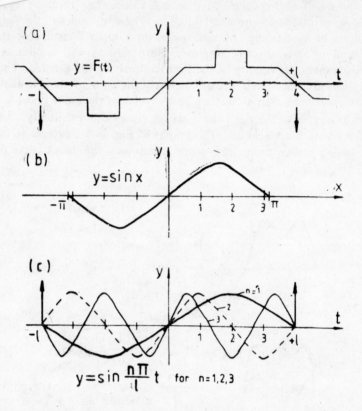

Fig. 14.1-2.   Linear Transformation of sin (nx) into sin $(n\pi t/l)$.

Fig. 14.1-2 (b) and (c). It allows to perform all calculations in time domain. The transformation of x into t is effected by the equation $x = m \cdot t$. The comparison of the graphs (a) and (b) shows for $x = \pi$, a corresponding $t = 1$. So $\pi = ml$ giving the transformation factor as $m = \pi/l$. So x and sin (nx) result as

$$x = \frac{\pi}{l} t$$

$$\sin (nx) = \sin \left(n\frac{\pi}{l} t \right) \qquad n\frac{\pi}{l} T = 2\pi \qquad T = \frac{2l}{n}$$

For each period T, the argument of the sine increases by 2π. The order of the harmonics is given by $n = 1, 2, 3,\dots$. So the second harmonic has half the period duration of the first harmonic $(T_2 = T_1/2)$, the third harmonic a third of it $(T_3 = T_1/3)$, etc., see (c).

14.2 FUNDAMENTALS OF FOURIER-ANALYSIS

Non sinusoidal, periodically alternating quantities are signified by the equation

$$Y\left(\frac{\pi}{l}\,t\right) = Y\left(\frac{\pi}{l}\,t + \frac{\pi}{l}\,T\right).$$

Such a function is depicted in Fig. 14.2-1. Fourier has proved that it may be composed of a trigonometric series of circular function as

$$Y = a_0 + F_1 \sin\left(\frac{\pi}{l}\,t + \varphi_1\right) + F_2 \sin\left(2\,\frac{\pi}{l}\,t + \varphi_2\right) + \dots$$

$$+ F_n \sin\left(n\,\frac{\pi}{l}\,t + \varphi_n\right) + \dots$$

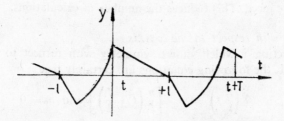

Fig. 14.2-1. Example of a Non-Symmetric Function.

Using the formula for $\sin(a+b) = \sin b \cos a + \cos b \sin a$, the n-th harmonic can be written as

$$F_n \sin\left(n\,\frac{\pi}{l}\,t + \varphi_n\right) = \underbrace{F_n \sin \varphi_n}_{a_n} \cdot \cos\left(n\,\frac{\pi}{l}\,t\right) + \underbrace{F_n \cos \varphi_n}_{b_n} \cdot \sin\left(n\,\frac{\pi}{l}\,t\right)$$

As $\cos(n\pi t/1)$ is certainly phase shifted towards $\sin(n\pi t/1)$ by 90° the law of **PYTHAGORAS** allows to determine the amplitude F_n of the n-th harmonic (having a circular frequency $\omega_n = n\pi/1$) as

$$F_n = \sqrt{a_n^2 + b_n^2}\,.$$

Using the previous equation the complete Fourier series may be expressed as

$$Y = a_0 + \sum_{n=1}^{\infty} a_n \cos\left(n\,\frac{\pi}{l}\,t\right) + \sum_{n=1}^{\infty} b_n \sin\left(n\,\frac{\pi}{l}\,t\right)$$

$$a_0 = \frac{1}{2l} \int_{-l}^{+l} Y(t)\, dt$$

$$a_n = \frac{1}{l} \int_{-l}^{+l} Y(t) \cdot \cos \left(n \frac{\pi}{l} t \right) dt$$

$$b_n = \frac{1}{l} \int_{-l}^{+l} Y(t) \cdot \sin \left(n \frac{\pi}{l} t \right) dt$$

The series consists of the mean value a_0 of the function $y(t)$ and a sum of cosine and sine functions having the coefficients a_n and b_n which constitute the amplitude F_n of the nth harmonic.

14.3 SYMMETRICAL FEATURES OF DIFFERENT SIGNAL SHAPES

In case the origin of the coordinates can be chosen in such a way that y shows certain symmetrical features, the number of integrals may be considerably reduced. This reduces the number of calculations.

Symmetry with respect to the x-Axis
For a function y which shows symmetry with respect to the x-axis, see Fig. 14.3-1, the following equations are certainly true

$$Y\left(\frac{\pi}{l} t \right) = - Y\left[\frac{\pi}{l} \left(t + \frac{T}{2} \right) \right] \text{ and } a_0 = 0$$

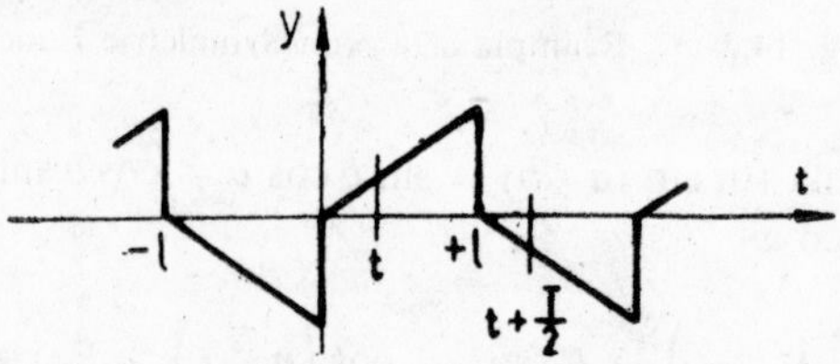

Fig. 14.3-1 Example of a Pulse Shape Symmetrical to *x*-Axis.

So the Fourier series may be written for both arguments as

$$Y\left(\frac{\pi}{l} t \right) = \sum_{n=1}^{\infty} a_n \cos \left(n \frac{\pi}{l} t \right) + \sum_{n=1}^{\infty} b_n \sin \left(n \frac{\pi}{l} t \right)$$

$$Y\left[\frac{\pi}{l} \left(t + \frac{T}{2} \right) \right] = \sum_{n=1}^{\infty} a_n \cos \left(n \frac{\pi}{l} t + \pi \right) + \sum_{n=1}^{\infty} b_n \sin \left(n \frac{\pi}{l} t + \pi \right)$$

Using $\cos (nx + \pi) = (-1)^n \cos (nx)$; $\quad \sin (nx + \pi) = (-1)^n \sin (nx)$
we obtain

$$Y\left[\frac{\pi}{l} \left(t + \frac{T}{2} \right) \right] = \sum_{n=1}^{\infty} (-1)^n a_n \cos \left(n \frac{\pi}{l} t \right) + \sum_{n=1}^{\infty} (-1)^n b_n \sin \left(n \frac{\pi}{l} t \right)$$

Due to the symmetry with respect to the x-axis, the negative value of the last equation should be equal to $y(\pi t / l)$. This is possible only for $n = 1, 3,$

5, ... So it is enough to calculate just the a_n and b_n values for the odd harmonics. Only the amplitudes F_1, F_3, F_5 exist. All others are zero. The mean value a_0 is zero, too.

Symmetry with Respect to the y-Axis

For a function y which shows symmetry with respect to the y-axis, see Fig. 14.3-2, the following statements are certainly true

$$Y\left(\frac{\pi}{l}\, t\right) = Y\left(-\frac{\pi}{l}\, t\right) \qquad a_0 \neq 0$$

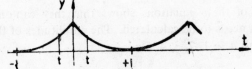

Fig. 14.3-2. Example of a Pulse Shape Symmetrical to y-Axis (Even Function).

So the Fourier series may be written for the positive and negative arguments as

$$Y\left(\frac{\pi}{l}\, t\right) = a_0 + \sum_{n=1}^{\infty} a_n \cos\left(n\frac{\pi}{l}\, t\right) + \sum_{n=1}^{\infty} b_n \sin\left(n\frac{\pi}{l}\, t\right)$$

$$Y\left(-\frac{\pi}{l}\, t\right) = a_0 + \sum_{n=1}^{\infty} a_n \cos\left(n\frac{\pi}{l}\, t\right) + \sum_{n=1}^{\infty} - b_n \sin\left(n\frac{\pi}{l}\, t\right)$$

In the last line it was considered that $\cos(-x) = \cos x$ and $\sin(-x) = -\sin x$.

The direct comparison of the two equations for $y(\pi t/l)$ and $y(-\pi t/l)$ shows that they can only be equal if $b_n = 0$. So b_n need not be calculated for such a function of y. The amplitudes of the harmonics are $F_n = a_n$ for all $n = 1, 2, 3, 4, \ldots$.

Symmetry with Respect to Zero on Both Axis (Centric Symmetry)

For a function which shows symmetry with respect to zero on both axis (centric symmetry) certainly the following statements can be read from Fig. 14.3-3:

$$Y\left(\frac{\pi}{l}\, t\right) = -Y\left(-\frac{\pi}{l}\, t\right) \qquad a_0 = 0$$

They allow to write the Fourier series as follows:

$$Y\left(\frac{\pi}{l}\, t\right) = a_0 + \sum_{n=1}^{\infty} a_n \cos\left(n\frac{\pi}{l}\, t\right) + \sum_{n=1}^{\infty} b_n \sin\left(n\frac{\pi}{l}\, t\right)$$

$$-Y\left(-\frac{\pi}{l}\, t\right) = -a_0 + \sum_{n=1}^{\infty} - a_n \cos\left(n\frac{\pi}{l}\, t\right) + \sum_{n=1}^{\infty} + b_n \sin\left(n\frac{\pi}{l}\, t\right)$$

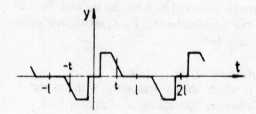

Fig. 14.3-3   Example of a Curve with Centric Symmetry
(Odd Function).

The comparison of these equations shows that they can only be equal if $a_n = 0$. So no a_n needs to be calculated. The amplitudes of the harmonics are $F_n = b_n$ for all $n = 1, 2, 3, 4$, etc.

14.4   EXAMPLE FOR CALCULATIONS: PERIODIC RECTANGULAR PULSES (SUPERIMPOSITION OF THE HARMONICS AND AMPLITUDE SPECTRUM)

The pulses have an amplitude A, a period duration of $T = 2l$ and a pulse ratio of $p = 2c/2l = c/l$ occasionally also referred to as mark-space-ratio. The function y is an even one, or in other words, it is symmetrical to the

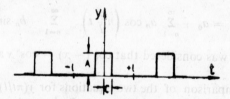

Fig. 14.4-1.   Periodic Square Pulses.

y-axis. So only the mean value a_0 and the coefficients of a_n of the cosine terms need be determined. They are the amplitudes F_n of the harmonics. All coefficients of b_n are zero.

$$a_0 = \frac{1}{2l} \int_{-l}^{+l} Y(t)\, dt = \frac{2}{2l} \int_{0}^{+c} A\, dt = A \cdot \frac{c}{l} = A \cdot p$$

$$a_n = \frac{1}{l} \int_{-l}^{+l} Y(t) \cdot \cos\left(n\frac{\pi}{l} t\right) dt$$

$$= \frac{2}{l} \left[\int_{0}^{c} A \cos\left(n\frac{\pi}{l} t\right) dt + \int_{c}^{l} 0 \cdot \cos\left(n\frac{\pi}{l} t\right) dt \right]$$

$$= \frac{2}{l} A \frac{l}{n\pi} \sin\left(n\frac{\pi}{l} t\right)\Big|_0^c$$

$$= \frac{2A}{n\pi} \sin\left(n\frac{\pi}{l} c\right) = \frac{2A}{n\pi} \sin(n\pi p).$$

Ex. In case we choose the amplitude $A = 1$ and the pulse ratio as $p = \frac{1}{4}$ the pulse function y may be written completely as

$$Y = \frac{1}{4} + \sum_{n=1}^{\infty} \underbrace{\frac{2}{n\pi} \sin\overbrace{\left(\frac{n\pi}{4}\right)}^{n\pi p}}_{a_n} \cdot \underbrace{\cos\left(n\frac{\pi}{l} t\right)}_{\omega_n}$$

Putting n as integer figures of $n = 1, 2, ..., 8$ we get the amplitudes of the first eight harmonics for different pulse ratios p, see table and Fig. 14.4-2.

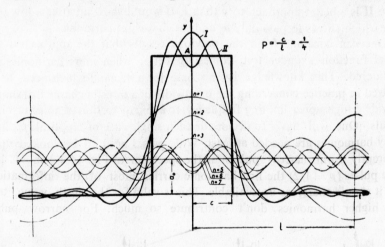

Fig. 14.4-2. Superimposition of the First Harmonics of Periodic Block Pulses.

p	a_0	a_1	a_2	a_3	a_4	a_5	a_6	a_7	a_8	$a_0 + \sum_{n=1}^{\nu} a_n = A'$	
1/2	0.50	+0.64	0	−0.21	0	+0.13	0	−0.09	0	$\nu = 2$ $\boxed{1.14}$	
1/4	0.25	+0.45	+0.32	+0.15	0	−0.09	−0.11	−0.06	0	$\nu = 4$ $\boxed{1.17}$	$>A = 1$
1/8	0.12	+0.24	+0.22	+0.20	+0.16	+0.12	+0.07	+0.03	0	$\nu = 8$ $\boxed{1.16}$	

For $p = 1/4$ the first three harmonics are positive, the fourth is zero, and the next three are negative. (This is quite similar for other pulse ratios P). These characteristics are essential for the composition of the block pulse from its harmonic components.

The mean value a_0 is the reference for all harmonics around which they oscillate with different amplitudes a_n and different frequencies f_n. Curve I of Fig. 14.4-2 is composed from a_0 and the first three harmonics. They are all cosine functions starting with their positive amplitudes at the time instant $t = 0$. Two and three period durations of the second and third harmonics can be found respectively within one period duration of the first harmonic being $T = 2l$. The sum of their instantaneous values plus a_0 produces curve I. It would appear after a low pass which blocks all harmonics higher than the fourth one. Curve II takes also the next group of harmonics into consideration. They all prove to have negative amplitudes in common ($n = 5, 6, 7$). Five, six and seven period durations of them fit into the one of the first harmonic. All of them start negatively which causes a saddle of curve II for $t = 0$: The sum of all negative amplitudes taken away from the maximum value of curve I produces the lowest point of the saddle. For time instants, near the sudden change of the original pulse function y the slope of curve II is increased compared to I. On the whole, curve II is a better approach to y than I. II would show up after a low pass filter which allows to pass only a_0 and the first eight harmonics.

One can conclude from the previous results that the approach to the actual function y under test, comes up better when more harmonics are considered. This knowledge is the base for a demand which needs to be realized in practice concerning the bandwidth of a transfer channel: It should provide a high upper limiting frequency, to avoid distortion of nonsinusoidal signals. One will have to decide for each single case of application, how many harmonics are needed at least. Wide pulses need to be looked upon differently from narrow pulses as the spectrum of Fig. 14.4-3 shows. For wide pulses ($p = 1/2$) the first harmonic carries most of the information, and it should therefore be available. This is valid for the mean value, too. But higher harmonics don't contribute so much. For narrow pulses

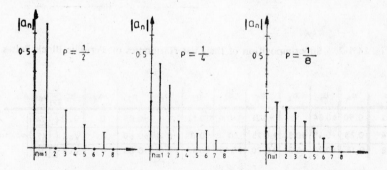

Fig. 14.4-3.   Amplitude Spectrum of Periodic Block Pulses for Different Pulse Ratio p.

($p = 1/8$) this is different however: Higher harmonics carry a lot of information and hence need to be considered. But the mean value a_0 is of low importance compared to the harmonics. So a pure AC transfer channel might serve all needs even though it does not allow to pass a_0. The technical realization will certainly take into account the economic dimensions, too. Though a narrow band width of a transfer channel might distort a signal considerably it still might pose the cheapest solution, though the extent of distortion cannot be accepted. But by employing triggers the signal could be reconditioned to its original shape, after having passed the transfer line, fitting all needs.

Quite often the question for the largest amplitude a_n arises. It is easily answered: As $\sin(n\pi p)$ can reach at the utmost the value of one only, the amplitude of the n-th harmonic cannot exceed the value of $2A/n\pi$. The first harmonic provides the biggest amplitude $a_1 = 2A/\pi$.

Effect of Low Pass Behaviour of the Transfer Path

Ex. A practical problem is certainly of high importance and relevance. It is the question of the highest output voltage after a low pass filter. The previous table allows to answer it, using the last column. If the filter allows to pass the DC component and the first two harmonics, it produces for a pulse ratio $p = 1/2$ and a pulse amplitude of $A = 1$ V a maximum output voltage $A' = 1.14$ V. For $p = 1/8$ and a bandwidth up to the eighth harmonic $A' = 1.16$ V. Obviously the output of the filter produces a maximum output voltage A' higher than the pulse voltage A. This fact might pose a danger to sensitive semiconductor components such as field effect transistors. They need to be selected for A' (instead of A) this way avoiding the risk of damaging them.

14.5 PERIODIC SAW TOOTH AND PARABOLIC PULSES

The function y of Fig. 14.5-1 provides centric symmetry. So only sine terms need to be calculated. The a_n are zero and a_0, too. For the straight lines of the saw tooth y may be written as

$$Y = \begin{cases} \dfrac{A}{c} \cdot t & \text{for } 0 \leqslant t \leqslant c \\[2mm] \dfrac{A(l-t)}{l-c} & \text{for } c \leqslant t \leqslant l \end{cases}$$

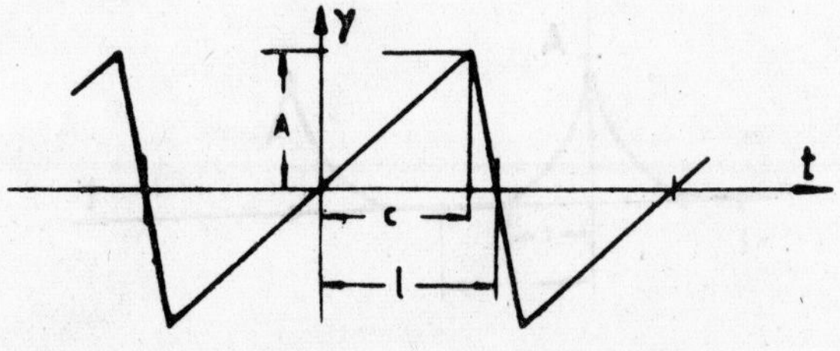

Fig. 14.5-1. Periodic Saw Tooth.

For the amplitudes b_n the following integrations need to be effected as

$$b_n = \frac{2}{l} \int\limits_0^c Y \cdot \sin\left(n\frac{\pi}{l}t\right) dt$$

$$= \frac{2}{l} \left[\frac{A}{c} \int\limits_0^c t \sin\left(n\frac{\pi}{l}t\right) dt + \frac{Al}{l-c} \int\limits_c^l \sin\left(n\frac{\pi}{l}t\right) dt \right.$$

$$\left. - \frac{A}{l-c} \int\limits_c^l t \cdot \sin\left(n\frac{\pi}{l}t\right) dt \right]$$

$$= \frac{2A \sin(n\pi p)}{n^2\pi^2 p(1-p)} \qquad (n = 1, 2, 3, \ldots\ldots)$$

The letter p gives the pulse ratio as $p = c/l$. A certain case of practical interest occurs for $c = l$ which makes $p = 1$. For this genuine type of a saw tooth, b_n becomes indeterminable using the last equation. It adopts the form of $0/0$. But approaching the limit for $p \to 1$ we obtain the result of the amplitudes b_n' as

$$b_n' = \lim_{p \to 1} b_n = -\frac{2A}{n\pi} \cos(n\pi) = \begin{cases} +\dfrac{2A}{n\pi} & (n = 1, 3, 5, \ldots) \\[2mm] -\dfrac{2A}{n\pi} & (n = 2, 4, 6, \ldots) \end{cases}$$

A detailed view would show the paramount importance of the first harmonic. Higher harmonics are not such substantial for the superimposition.

Parabolic Pulses

For the function y of Fig. 14.5-2 the parabolic branches are analytically described as

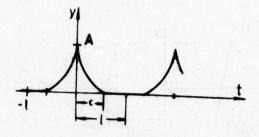

Fig. 14.5-2.   Parabola Pulses.

$$Y = \begin{cases} \dfrac{A}{c^2}(t-c)^2 & \text{for } 0 \leqslant t \leqslant c \\ \\ 0 & \text{for } c \leqslant t \leqslant l \end{cases}$$

Being symmetrical with respect to the y-axis only a_0 and a_n need to be calculated. Again $p = c/l$. The results are

$$a_0 = \frac{Ap}{3} \quad \text{and} \quad a_n = \frac{4A}{n^2\pi^2 p}\left(1 - \frac{\sin(n\pi p)}{n\pi p}\right); (n = 1, 2, 3, \ldots)$$

14.6 USEFUL INTEGRALS

A review of the three selected functions y, as rectangular saw tooth or parabolic pulses shows that they are composed from straight lines parallel to the y- and x-axis or from sloped linear functions, or from parabolic functions respectively. More complex functions may usually be composed from pieces of these investigated functions. For the determination of the Fourier coefficients the following integrals may appear. Using their solutions any Fourier analysis can be performed analytically.

$$\int \sin\left(n\frac{\pi}{l}t\right) dt = -\frac{l}{n\pi}\cos\left(n\frac{\pi}{l}t\right)$$

$$\int \cos\left(n\frac{\pi}{l}t\right) dt = +\frac{l}{n\pi}\sin\left(n\frac{\pi}{l}t\right)$$

$$\int t \cdot \sin\left(n\frac{\pi}{l}t\right) dt = -\frac{lt}{n\pi}\cos\left(n\frac{\pi}{l}t\right) + \frac{l^2}{n^2\pi^2}\sin\left(n\frac{\pi}{l}t\right)$$

$$\int t \cdot \cos\left(n\frac{\pi}{l}t\right) dt = +\frac{lt}{n\pi}\sin\left(n\frac{\pi}{l}t\right) + \frac{l^2}{n^2\pi^2}\cos\left(n\frac{\pi}{l}t\right)$$

$$\int t^2 \cdot \sin\left(n\frac{\pi}{l}t\right) dt = -\frac{lt^2}{n\pi}\cos\left(n\frac{\pi}{l}t\right) + \frac{2l^2 t}{n^2\pi^2}\sin\left(n\frac{\pi}{l}t\right)$$
$$+ \frac{2l^3}{n^3\pi^3}\cos\left(n\frac{\pi}{l}t\right)$$

$$\int t^2 \cdot \cos\left(n\frac{\pi}{l}t\right) dt = +\frac{lt^2}{n\pi}\sin\left(n\frac{\pi}{l}t\right) + \frac{2l^2 t}{n^2\pi^2}\cos\left(n\frac{\pi}{l}t\right)$$
$$\times \frac{2l^3}{n^3\pi^3}\cos\left(n\frac{\pi}{l}t\right)$$

14.7 PRACTICAL HARMONIC ANALYSIS USING A WATTMETER

A non sinusoidal current $I(t)$ passing a choke, being fed from a sine voltage from the line should be investigated for its harmonics I_n. All practical needs may be met using a simple watt meter, see Fig. 14.7-1. The current $I(t)$ can be written as sum of sine terms as it was done in chapter 14.2. The mean value is zero now. This way of writing $I(t)$ facilitates the following analysis :

$$I(t) = \sum_{n=1}^{\infty} \hat{I}_n \sin n(\omega_I t + \varphi_{I,n})$$

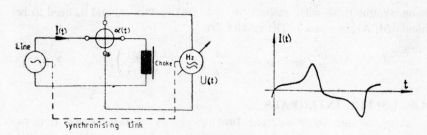

Fig. 14.7-1. Distortion Analysis **Circuit** for Investigating a Non-Sinusoidal Current **using** a Watt **Meter**.

This current passes the current path of the watt meter. A voltage $U(t)$ from a sine generator which allows alteration in its circular frequency ω_u is applied to the voltage path.

$$U(t) = \hat{U} \sin (\omega_u t + \varphi_u)$$

The phases $n \cdot \varphi_{I, n}$ of the harmonics (I_n) and the one φ_u of the voltage $U(t)$ are the angles towards an $(\omega t) = 0$ which might be freely chosen. But only the phaseshifts $(\varphi_u - n \cdot \varphi_{I, n}) = \varphi_n$ of the current harmonics towards the voltage $U(t)$ are of interest.

14.7.1 Amplitude Determination

The amplitudes $\hat{I}_n$ of the different harmonics should be determined first. A wattmeter allows to do that, as it produces an electric torque which is proportional to the instantaneous product of

$$P(t) = U(t) \cdot I(t)$$

$$= \hat{U} \sum_{n=1}^{\infty} \hat{I}_n \underbrace{\sin (\omega_u t + \varphi_u)}_{a} \cdot \underbrace{\sin n(\omega_I t + \varphi_{I, n})}_{b}$$

$$\sin a \sin b = \tfrac{1}{2} [\cos (a - b) - \cos (a + b)]$$

$$P(t) = \tfrac{1}{2} \hat{U} \sum_{n=1}^{\infty} \hat{I}_n \{\cos [(\omega_u - n\omega_I) t + \varphi_u - n\varphi_{I, n}] - \cos [(\omega_u + n\omega_I)t + \varphi_u + n\varphi_{I, n}]\}$$

Due to the torque inertia of the moving coil, normally the indication α of the pointer would be zero, for it effects an integration of $P(t)$ which is a pure alternating quantity. But if ω_u is chosen close to $n\omega_I$ especially so close that the difference of the circular frequencies $(\omega_u - n\omega_I)$ for a certain harmonic n becomes less than 1 Hz, the pointer might follow these slow changes and indicate a changing deflection $\alpha(t)$. The second term of the difference will not be considered for this case. It changes very fast with the circular frequency of $(\omega_u + n\omega_I)$ and will be integrated to zero. So the instantaneous indication of $\alpha(t)$ corresponds directly to the term containing $(\omega_u - n\omega_I)\, t$.

A maximum reading $\hat{\alpha}_n$ may be taken at the time instant for which $\cos\left[(\omega_u - n\omega_I)t + \varphi_u - n\varphi_{I, n}\right] = 1$:

$$\hat{\alpha}_n = c_w \,\hat{U}\,\hat{I}_n$$

$$\hat{I}_n = \frac{\hat{\alpha}_n}{c_w\,\hat{U}}$$

The wattmeter constant c_w needs to be known in scale divisions per watt— $\hat{I}_n$ is, for instance, the amplitude of the first harmonic of the current $I(t)$ under test if the frequency of the sine voltage generator was chosen close to the one of the first harmonic. Other harmonics are found by tuning the frequency ω_u to higher values until the pointer responds again with slow motions to allow the reading of another $\hat{\alpha}_n$.

14.7.2 Phase Determination
The circuit of Fig. 14.7-1 allows to determine the phase angles φ_n of the harmonics, too. For this purpose the circular frequency ω_u needs exactly to be set equal to $n\omega_I$. The difference of the circular frequencies will be zero then. The first cosine in $P(t)$ contains a constant argument $(\varphi_u - n\cdot\varphi_{I, n}) = \varphi_n$ being the phase angle of interest which does not depend on time. (The second cosine alternates very fast and is therefore integrated to zero). A certain constant rea ing α_n is obtainable as

$$\alpha_n = c_w \,\hat{U}\,\hat{I}_n\,\cos\underbrace{(\varphi_u - n\varphi_{I, n})}_{\varphi_n}$$

$$\varphi_n = \arccos\frac{\alpha_n}{c_w\,\hat{U}\,\hat{I}_n}$$

$\hat{I}_n$ was already determined.

Practical Hints
To make sure that ω_u equals $n\cdot\omega_I$ totally, we have to synchronize the $U(t)$ generator by the line. The synchronization needs to be ensured carefully.

Choose a wattmeter which provides a light torque inertia. This ensures that it can follow the difference frequency even if it is not too small which cases the tuning.

Current and voltage paths should be able to take high overloads because higher harmonics have usually smaller amplitudes, resulting in smaller indication angles. But all other harmonics as well as the indicated one still load the instrument to their part of power they contain.

The power meter should have its zero in the middle of the scale. Thus a positive and a negative reading of $\hat{\alpha}_n$ can be taken. By calculating the mean value of the two readings a zero shift of the pointer is eliminated.

The current path of a wattmeter allows usually at least a frequency of 1000 Hz. So a Fourier analysis for currents from the line of 50 Hz for the first harmonic can be effected, up to the 20th harmonic, still providing sufficient accuracy.

14.8   ELECTRONIC DISTORTION ANALYSER

Fig. 14.8-1 shows the block diagram of an electronic distortion analyser, realizing the same action principle as for the previously described method using a wattmeter.

The measurement voltage U_M of nonsinusoidal nature is to be analyzed for its harmonics of the order $n = 1, 2, 3,\ldots$, etc. A mixer modulates the frequencies to be investigated $(n\omega)$ and the tunable carrier frequency (Ω).

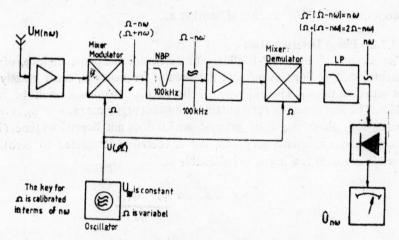

Fig. 14.8-1.   Electronic Distortion Analyser.

Their multiplication produces a difference frequency of $(\Omega - n\omega)$ and a sum frequency of $(\Omega + n\omega)$, as shown in chapter 14.7.1. A narrow band pass filter (NBP) is exactly tuned to a constant intermediate frequency, usually being 100 kHz, which it allows to pass excludingly. In order to obtain an output signal of difference frequency, Ω is tuned to such a value which produces $(\Omega - n\omega)/2\pi = 100$ kHz. This again is mixed with Ω producing another difference frequency of $\{\Omega - [\Omega - n\omega]\}/2\pi = n\omega/2\pi$. This is separated from the frequency sum with the help of a low pass (LP) producing an output signal with the frequency $(n\omega)/2\pi$ to be investigated. After having rectified it, its amplitude $\hat{u}_{n\omega}$ may directly be indicated.

If U_M is the input voltage of an aerial of a radio receiver containing the transmission frequencies of different broadcasting stations (e.g. $n_1\omega$ for London, $n_2\omega$ for Paris, and $n_3\omega$ for Dar es Salaam) Ω needs to be set to certain $\Omega_1, \Omega_2, \Omega_3$ to select these stations. A successful selection can be indicated by the meter. Usually the position of the selection key for Ω is calibrated in terms of the nominal transmission frequencies $n \cdot \omega$ of the different stations, thus facilitating their search. But actually Ω is 100 kHz higher than the $n\omega$ of a certain station.

14.9 ANALYSIS OF A PHASE CUT CURRENT

Ex. A phase cut current I as it passes a resistive load R_L which is controlled by a triac controller will contain harmonics $I_1, I_3, I_5,$ etc., see Fig. 14.9-1. Their amount depends on the firing angle δ. Employing the previous methods

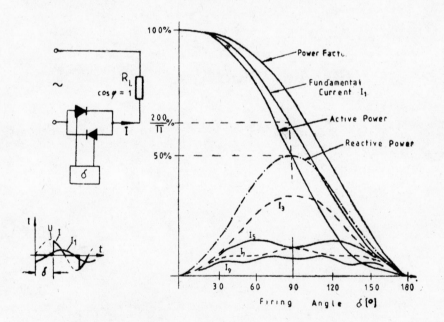

Fig. 14.9-1. Analysis of a Phase Cut Sine Current through an Active Load.

the Fourier analysis of the current I may be performed for different firing angles δ. The results are depicted in the same figure.

Only the fundamental current produces active power. All overtones cause reactive power. The power factor is an appropriate measure to signify this (stunning) fact. All suppliers of electrical energy limit the permissible lowest power factor to a certain value, this way avoiding high line currents which are not entirely used for the active energy transfer. In order to meet the power factor condition the consumer needs to compensate at least a part of the reactive power, even in case of pure ohmic load which is in need of reactive power, too, if the current is phase cut.

15 Instrument Transformers

Instrument transformers serve measurement functions. They are available as voltage or current devices, preferably to step down voltages and currents to ranges which allow an easy handling or processing.

15.1 GENERAL USE OF VOLTAGE AND CURRENT TRANSFORMERS

Their main application can be found in the field of power and high tension technics for the following reasons:

Ex. 1. Voltage transformers are needed to step down the high voltage to be measured (say $U_m = 100$ kV) to a low handy one (say $U_v = 100$ V) to avoid unnecessary measurement losses. If, for instance, a soft iron voltmeter is used which produces full scale deflection (f.s.d.) for $I_v = 10$ mA passing its coil the necessary series resistor R_S would consume a considerable power.

$$P = (U_m - U_v) \cdot I_v = (100000 \text{ V} - 100 \text{ V}) \cdot 10 \text{ mA} \approx 1 \text{ kW}$$

This power loss P can be avoided using a transformer.

2. A transformer facilitates the separation from the high tension line, thus providing safety for the operator who handles the equipment.

3. Current measurements could also be effected using shunts. But its resistance in combination with the instrument reactance produces phase errors, which may become considerably high for power measurements.

4. Even more important is the fact that the output current of a current transformer depends only on the load current to be measured. So the internal impedance of the ammeter is not important within reasonable limits, a fact which provides more flexibility for the measurement arrangement if a transformer is employed.

15.2 VOLTAGE INSTRUMEMT TRANSFORMER (EQUIVALENT CIRCUIT AND PHASOR DIAGRAM)

The equivalent circuit can be developed following the depictions (a) to (d) of Fig. 15.2-1. In practice the primary (N_1) and secondary (N_2) windings generate the main flux ϕ_M as well as the stray fluxes $\phi_{\sigma 1}$ and $\phi_{\sigma 2}$. $\phi_{\sigma 1}$ and

ϕ_{σ_2} are not linked to the other coil, see Fig. 15.2-1(a). This fact can be accounted for by employing stray reactances X_{σ_1}, and X_{σ_2}, see Fig. (b). The resistances of the windings R_1 and R_2 are considered as lumped resistors.

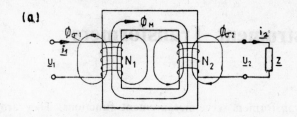

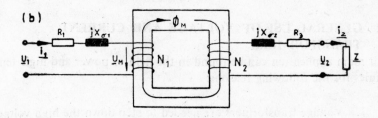

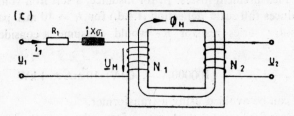

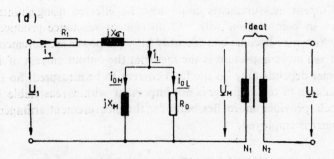

Fig. 15.2-1. Development of the Equivalent Circuit of the Voltage Instrument Transformer.

The transformer shows ideal performance now. The main flux ϕ_M links both windings N_1 and N_2. ϕ_M is caused by the main voltage U_M. U_M is obtainable from U_1 after taking away the voltage drops across R_1 and $X_{\sigma 1}$, which originated from the primary current i_1.

An instrument transformer for measurements of voltag s is usually not loaded. A voltmeter draws only a negligible current: $i_2 \approx 0$. So voltage drops across the secondary stray reactance $X_{\sigma 2}$ and the winding resistance R_2 are neglected, as shown in Fig. (c).

The primary current i_1 serves to magnetize the core (i_{0M}) and to cover the hysteresis ($\sim f$) and the eddy current ($\sim f^2$) losses: i_{0L}. In this way the equivalent circuit changes into the depiction of Fig. (d). The main voltage U_M produces directly the secondary voltage $U_2 = U_M \cdot N_2/N_1$. So the need arises to provide negligible voltage drops across R_1 and $X_{\sigma 1}$ in order to obtain U_2 directly from U_1. U_1 should become (nearly) equal to the main voltage U_M. The following equations describe the transformer behaviour, and at the same time they form the base for the phasor diagram of Fig. 15.2-2 (a).

$$U_1 = U_{R1} + U_{x\sigma 1} + U_M \qquad\qquad i_1 = i_{0M} + i_{0L}$$

$$= i_1 (R_1 + jX_{\sigma 1}) + U_M$$

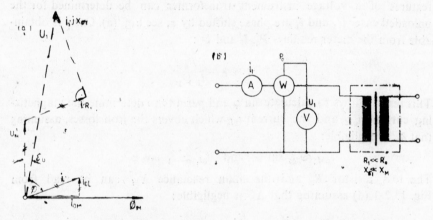

Fig. 15.2.2. **Phasor Diagram** of a Voltage Instrument Transformer (a) and the Test Circuit (b).

The magnetizing current i_{0M} is in phase with the main flux ϕ_M being generated by the main reactance X_M. The current i_{0L} passes through the loss resistor R_0 (hysteresis and eddies) being phase shifted by exactly 90°. The main voltage U_M is found across R_0, so it is in phase with i_{0L}. The currents i_{0M} and i_{0L} compose the primary current i_1 which causes the voltage drop $i_1 \cdot R_1$ (being in phase with i_1), and the one across the stray reactance $i_1 \cdot jX_{\sigma 1}$ (being exactly phase shifted by 90°). The sum of both voltages plus U_M makes up U_1.

According to Fig. 15.2-2 (a) neither the magnitude of U_1 equals the magnitude of U_M (magnitude error $\Delta U = |U_M| - |U_1|$) nor are they in the same phase (angular error δ_u). In order to avoid non-linearity due to the non-linear magnetizing curve, the induction $B_M = \phi_M/A$ should not be high. A core of large cross-section area A satisfies this condition. The core material needs to be of high permeability in order to obtain an adequate main flux ϕ_M for a low magnetizing current i_{0M}.

The magnitude error can be made small, with the help of a toroidal core which gives very little stray flux, thus ensuring a low $X_{\sigma1}$. By employing thick copper wire for the primary windings R_1 is designed small, in such a way reducing the angular error. The current i_{0L} which covers the iron losses will be small for a core material which is well laminated to avoid eddy currents, and it should provide a very low area of the hysteresis loop to avoid hysteresis losses.

Measurement of Transformer Features
For an instrument transformer the previously described features are sufficiently realized. They include the following statements:

$$R_1 \ll R_0 \quad \text{and} \quad X_{\sigma1} \ll X_M$$

Employing an ammeter, a voltmeter and a wattmeter (see Fig. 15.2-2 b) the features of a voltage instrument transformer can be determined for the unloaded case. U_1 and i_1 are phase shifted by φ, see Fig. (a). $\cos \varphi$ is obtainable from the meter readings P_0, i_1 and U_1:

$$\cos \varphi = \frac{P_0}{i_1 \, U_1}$$

This result allows to calculate $\sin \varphi$ and permits to determine the magnetizing current i_{0M} and the current i_{0L} which covers the iron losses, assuming that δ_u is negligible:

$$i_{0M} \approx i_1 \sin \varphi \quad \text{and} \quad i_{0L} \approx i_1 \cdot \cos \varphi$$

The loss resistor R_0 and the main reactance X_M can be read from Fig. 15.2-1 (d) assuming that ΔU is negligible:

$$R_0 = \frac{U_M}{i_{0L}} \approx \frac{U_1}{i_1 \cdot \cos \varphi}$$

$$X_M = \frac{U_M}{i_{0M}} \approx \frac{U_1}{i_1 \cdot \sin \varphi}$$

15.3 CURRENT TRANSFORMER
Current transformers are mainly used to step down high currents to low ones (e.g. 5 A) which can be measured easily. But in high tension lines the currents to be measured may be small already. In this case current transformers provide the necessary separation from the line for safety reasons.

Action Principle and Simplified Phasor Diagram
The current transformer is operated in its short circuited state. So the primary magnetic circulation becomes nearly equal to the secondary one.

The remaining difference represents the exciting magneto motive force (V_m) which is needed to set up the flux through the magnetic reluctance (i_{0M}) and to cover the hysteresis and eddy current losses (i_{0L}), see Fig. 15.3-1.

$$V_m = i_0 N_1 = (i_{0M} + i_{0L}) N_1 = i_1 N_1 - i_2 N_2$$

The error term $i_0 N_1$ should be zero. In this case only the turns ratio would determine the step down ratio of the currents:

$$\frac{i_1}{i_2} = \frac{N_2}{N_1}$$

However in reality, there is a transformation error which consists of a magnitude error $\Delta V_m = |i_2/N_2 - |i_1/N_1$ and a phase error δ_i.

In order to come near to the ideal relation, certain items need to be realized. They are summarized again as: 1. Low active losses are provided for by small area of the hysteresis loop of the core material. It needs to be laminated to avoid eddy currents. The copper losses of the coils are low because they are wound with thick wire to have low current density. 2. The

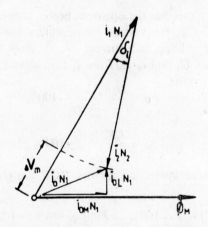

Fig. 15.3-1. Phasor Diagram of the Current Instrument Transformer.

magnetic stray losses are kept low due to the toroidal configuration of the core. 3. High permeability of the core provides for easy magnetizing of the core with low current i_{0M} which actually causes the magnetic circulation. 4. A wide cross section area of the core provides for good linearity due to low induction inside the iron.

Conclusions for Use

The load of a current transformer in action must not be changed, unless the output leads had been short circuited first. The unloaded case has to be avoided for the following reasons:

1. For an open output the secondary circulation becomes zero as there is no current ($i_2 = 0$). So the high primary circulation can magnetize the core up to its saturation: 1 to 2 Tesla. This causes high hysteresis losses which overheat the core, finally changing its magnetic features and eventually burning out the insulation of the coils. 2. Furthermore the far increased induction change dB/dt induces a high output voltage which might cause a breakdown of the winding insulation due to ionization effects. 3. The primary coil poses a choke of high inductivity towards the line which might change its features considerably.

15.4 RATED CAPABILITIES

The nominal transformer ratio is labelled as a fraction of nominal voltages or currents for voltage or current transformers respectively, such as

$$\frac{U_{1n}}{U_{2n}} = K_n = \frac{6000\text{V}}{100\text{V}}$$

$$\frac{I_{1n}}{I_{2n}} = K_n = \frac{100\text{A}}{5\text{A}}$$

The nominal power specifies the apparent power across the secondary load ($Z_n = 1/Y_n$) as $Y_n \cdot U_{2n}^2$ for voltage transformers and $Z_n \cdot I_{2n}^2$ for current transformers. The voltage and current errors show the percentage deviations of $U_2 \cdot K_n$ from U_1 and $I_2 \cdot K_n$ from I_1 respectively:

$$F_U = \frac{U_2 K_n - U_1}{U_1} \cdot 100\%$$

$$F_I = \frac{I_2 K_n - I_1}{I_1} \cdot 100\%$$

Occasionally the rated correction factors RCF are given instead of F_U or F_I being

$$\text{RCF}_{U,I} = 100\% - F_{U,I} \approx (99.2 \text{ to } 100.8)\%$$

The phase angle error δ_U or δ_I is usually less than 25 minutes of a degree. The overload factor n is defined only for current instrument transformers giving the ultimate multiple (e.g. $n < 5$) of the nominal primary current I_{1n} for nominal load causing an error F_I (e.g. $F_I = -10\%$).

15.5 LEAD MARKING AND TRANSFORMER CONNECTION

The lead marking and transformer connection follow certain norms to provide safety and to ensure equal conditions for repeated measurements. The primary leads are signified by capital letters, and the secondary ones by small letters, see Fig. 15.5-1. For line separating voltage transformers U, V and u, v are used. The auto-type employs X, U and x, u respectively. Current transformers are connected primarily at K, L and secondarily at *k, l.* The ground for voltage and current transformers should be applied to *v* and *k* respectively. For power measurements, these leads are common for the voltage and the current paths of the wattmeter. For safety reasons

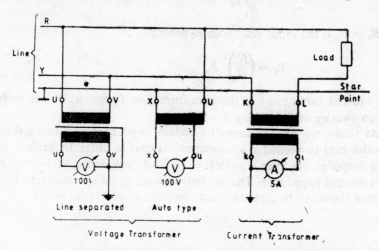

Fig. 15.5-1. Transformer Connections and Lead Markings.

the auto type voltage transformer should be employed only for phase voltage measurements. They are effected towards the star point, the potential of which can never be dangerously high. It is to be connected to *X, x.*

15.6 TRANSFORMER IMPEDANCE MATCHING

A voltage transformer may be used to convert an impedance, see Fig. 15.6-1. The output load R_0 appears as the input resistance R_i across the input leads 1 and 2. R_0 is converted into R_i with the help of the ratio of turns N_1 and N_2. The following derivation demonstrates this:

$$R_i = \frac{U_1}{i_1} \qquad \frac{U_1}{U_2} = \frac{N_1}{N_2} \qquad \frac{i_2}{i_1} = \frac{N_1}{N_2}$$

$$U_1 = \frac{N_1}{N_2} U_2 \qquad i_1 = \frac{N_2}{N_1} i_2$$

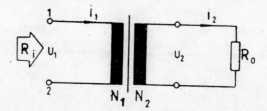

Fig. 15.6-1. Impedance Matching Employing a Voltage Transformer.

$$\therefore \quad R_i = \frac{\frac{N_1}{N_2} U_2}{\frac{N_2}{N_1} i_2} = \left(\frac{N_1}{N_2}\right)^2 \frac{U_2}{i_2}$$

As $R_0 = U_2/i_2$ the input resistance R_i becomes

$$R_i = \left(\frac{N_1}{N_2}\right)^2 R_0$$

The squared turns ratio is the transformation factor, which can be freely chosen meeting any matching condition.

At times, we have to convert a certain impedance Z_0 into another one Z_i which may be needed in telecommunications: In order to achieve maximum output power from a generator, its load impedance should be equal to its internal impedance. The maximum power should be available in cases of small signals to be measured, e.g. the voltage of a radio aerial.

16 Measurements of Magnetic Quantities

The magnetic field is basically defined by the quantities of induction B and field strength H. They are related to each other by the permeability of the material concerned, giving their ratio as $\mu = B/H$. The electric equivalents, as well as the basic rules, for designing a magnetic path have been dealt with in chapter 2.1.4. Now the measurements of these quantities are presented. At first we deal with the instruments needed for magnetic measurements, in chapter 16.1 and 16.2. Then we apply them as tools to sense a few features of ferromagnetic materials, in chapter 16.3. These instruments have been selected from lots of other instruments frequently in use.

16.1 SENSING OF INDUCTION B

The induction B of a magnetic field may be explored using a probe coil, being subject to the magnetic field of interest. This method makes use of the induction law signifying it as a dynamic measurement because a change of flux is needed. But static fields can be investigated, too, by making use of certain features of a thin semiconductor layer such as InAs or InSb. The first alloy may be designed as an 'active' device, known as Hall generator. It provides a voltage. InSb is applied to field plates providing a passive change of its resistance.

16.1.1 Probe Coil with Integrator

A coil being subject to a change of flux induces a voltage u_i. (Fig. 16.1.1-1) for constant coil area A.

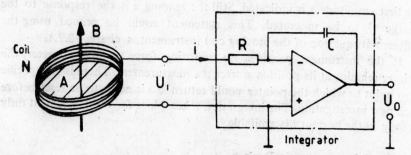

Fig. 16.1.-1.1   Measurement of Induction B with Probe Coil and Integrator
(R represents all loss resistors of the current circle for i).

$$U_i = - N\frac{d\phi}{dt} = - NA\,\frac{dB}{at}$$

For a large area A of the coil and lots of turns N the sensitivity of the device will be high. (If the coil is wound round an iron core it makes no sense to have A larger than that of the core.) The integrator provides an output voltage U_0 being proportional to the induction B to be measured:

$$U_0 = - \frac{1}{RC}\int U_i\,dt + U_s$$

$$= \frac{NA}{RC}\,B + U_s$$

If B is constant, as for a permanent magnetic field, it still can be measured by moving the coil from outside into it this way obtaining a change of flux. Independently from the way of introducing the coil to the spot of interest U_0 provides directly the information about the amount of B if the integrator had been set to $U_0\,(t = 0) = 0 = U_s$ before the measurement was started.

The charge Q being the time integral of i, provides B. So B may alternatively be measured with the help of a ballistic galvanometer. It is a moving coil instrument with a very small restoring torque inertia. Its deflection α (of the first amplitude) is proportional to Q:

$$\alpha = \frac{1}{C_B}\,Q = \frac{1}{C_B}\int i\,dt$$

The ballistic constant C_B of the instrument gives the charge needed to obtain one degree of pointer deflection.

The current i being limited by the total loss resistance R derives from the voltage u_i. So the deflection becomes

$$\alpha = \frac{1}{C_B R}\int U_i\,dt = \frac{1}{C_B R}\,NA \cdot B$$

For the ballistic galvanometer the current pulse may be finished even before the first amplitude α is indicated. Still the reading α is the response to the charge Q to be measured. This statement could be proved using the differential equation of the moving coil instrument, s. chapter 2.2.1.

If the instrument provides no restoring force at all (flux meter) the pointer remains at its position α after the measurement. There is no specific zero point to which the pointer could return. So it needs to be reset before each new investigation. The integration takes place for this instrument only as long as the current i is available.

16.1.2 Galvanomagnetic Devices
Sensors, making use of galvanomagnetic effects to measure the induction B,

are HALL generators and field plates. The B field interacts with the moving elementary charges (making up the current I_C) travelling through the device. The LORENTZ force deviates their traces changing the symmetrical field pattern within the sensor layer compared to the field as it was present for zero induction ($B = 0$).

HALL Generator

The control current I_C passes through the thin semiconductor layer with constant current density for zero induction. So the potential lines are parallel and straight. But applying an induction $B \neq 0$ causes them to incline and to bend, s. fig. 16.1.2-1. The transversely applied voltage connections are subject to different potentials providing the HALL voltage U_H as their difference:

$$U_H = R_H \frac{I_C}{d} B$$

The voltage U_H depends directly on the induction B to be measured and also on the control current I_C. The thickness d of the plate should be small to obtain a considerable voltage. The Hall constant $R_H = 100$ cm³/As for

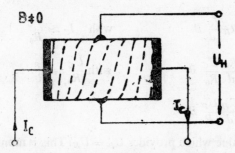

Fig. 16.1.2-1. Potential Lines in a HALL Generator Plate being Subject to an Induction Field B perpendicular to the surface.

InAs. It does not depend on temperature within a range of $(-40$ to $+150)$°C. Hall plates of 0.1 mm thickness and less are enhoused in flat ceramic enclosures. But they may be evaporated on substrates with $d < 0.01$ mm even. Commercial general purpose devices may dissipate $(0.1$ to $0 5)$ W, limiting the control current I_C to $(5$ to $250)$ mA depending on the type.

For measurements of strong fields $(B > 0.5$ Vs/m²) the Hall generator is usually directly positioned within the air gap of the object of interest, for example a magnet. U_H shows a tendency to increase more than linearly for strong increasing fields. Linearization can be effected with the help of a parallel resistor to U_H. The measurements of soft fields is facilitated by flux guiding cores which concentrate the field directly to the sensitive area of the sensor. If the induction is very low the Hall voltage U_H may be disturbed badly by thermoelectric voltages. It can be transmitted with the help of a transformer which rejects the thermoelectric DC voltage from further processing, see fig. 16.1.2-2. The induction B to be measured can be obtained

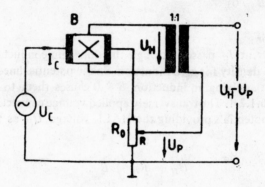

Fig. 16.1.2-2.   HALL Generator with Alternating Control Current I_C
(to reject DC voltage drifts of U_H)

from the potentiometer reading R, as shown by the following derivations. The Hall voltage was

$$U_H = R_H \frac{I_C}{d} B \qquad\qquad \text{with} \quad I_C \approx \frac{U_C}{R_0}$$

$$= \frac{R_H}{d} \frac{U_C}{R_0} B \qquad\qquad \text{with} \quad \frac{U_C}{R_0} = \frac{U_P}{R}$$

$$= \frac{R_H}{d} \frac{U_P}{R} B$$

Now, R is set to a value which provides $U_P = U_H$. This is monitored with a zero indicator across the output terminals: $U_P - U_H = 0$. In this case the ratio of the voltages becomes unit and B is obtained from the reading of R.

$$\frac{U_H}{U_P} = 1 = \frac{R_H}{d} \frac{B}{R}$$

$$B = \frac{d}{R_H} R$$

Current Meter with Hall Generator

As depicted in Fig. 16.1.2-3 the Hall Generator is positioned within the airgap of an iron core which couples the primary coil (N_1) with the secondary coil (N_2). The amplifier controls the secondary current I_2 to obtain a total flux $\phi = 0$. The magneto motive forces (mmf) compensate each other $(U_H = 0)$:

$$I_1N_1 - I_2N_2 = 0$$

$$I_2 = \frac{N_1}{N} I_1$$

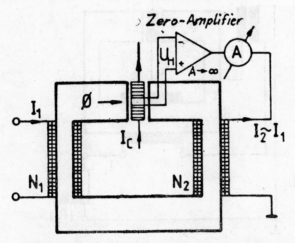

Fig. 16.1.2-3. Active Current Meter with Hall Generator.

If further processing of the measured current is desired, the ammeter may be replaced by any electronic circuitry. The 'active' instrument transformer of fig. 16.1.2-3 provides advantages over passive types:

— The iron core is free of any field even if the load of the secondary coil does not short circuit its terminals. Therefore, a non=linear core characteristic cannot interfere with the measurement and there are no iron losses. Also a transformation error (neither amplitudewise nor phasewise) cannot occur due to the lack of a magnetizing mmf.

— The transformer is capable to measure DC currents because Hall generators are static devices.

Power Meter with Hall Generator

A powermeter making use of a Hall generator may be of interest in case an electric output quantity should be available. The circuit of Fig. 16.1.2-4 provides an output voltage U_0 which is proportional to the power P_L of the load:

$$U_H = R_H \frac{I_C}{d} B \qquad \text{with } I_C = \frac{U_L/100}{R} \text{ and } B = C_B I_L$$

$$= R_H \frac{U_L}{100\,R} \frac{C_B}{d} I_L$$

$$U_0 = A U_H$$

$$= A R_H \frac{C_B}{100\,Rd} U_L I_L$$

$$= C \cdot P_L$$

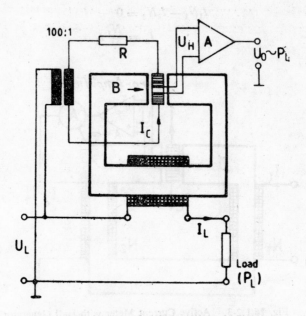

Fig. 16.1.2-4.   Power Meter with Hall Generator.

A change of B tends to change I_C because the resistance of the Hall plate changes with B. But still I_C needs to be constant. This can be provided for by choosing R sufficiently high. Consequently the ratio of the instrument transformer should not be too high.

The multiplication of $U_L \cdot I_L$ is basically effected likewise as for an electrodynamic power meter. So the range of applications of this circuit is the same as for the instruments of chapter 3.

Field Plates

Field plates make use of the Hall effect, too. But they are designed as two port devices. The induction B effects a resistance change due to the increased length of fieldlines along which the electrons travel. By implanting metal needles parallel to the component edges into the semiconductor material, the potential lines are forced to remain parallel and the electrons have to travel a considerably longer route between the outer electrodes, once an induction $B \neq 0$ is applied, see Fig. 16.1.2-5. Instead of going straight from electrode to electrode (as the charge carriers would do for $B = 0$) they are deviated for $B \neq 0$. They travel a far way from needle to needle on their wound path to the other electrode, as it is depicted for one field line (s. arrow) serving as representative of all others. The electrons follow deflected traces between the needles. The resistance R_F may increase twenty times for an induction of 1.2 Vs/m² compared to the resistance R_0 for zero induction, s. fig. 16.1.2-6. The sign of the field direction is not significant. But B is assumed to be perpendicular to the plate area. Around zero induction the sensitivity of the device is zero. However, subjecting the sensor to a superponed constant field, shifts the point of operation into the region of high sensivity. Now, even very small inductions B can be sensed, down to about $3 \ \mu Vs/m^2$ which is roughly 1/10 of the earth induction between the tropicals of capricorn and cancer.

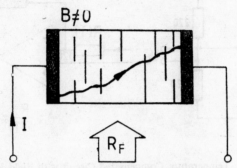

Fig. 16.1.2-5. Implanted conductive needles in a field plate to sustain parallel potential lines.

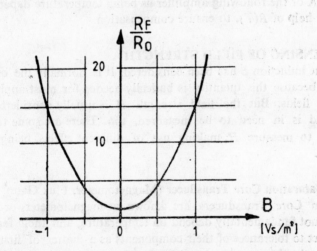

Fig. 16.1.2-6. Resistance Characteristic $R = f(B)$ of a Field Plate.

Field plates are badly dependent on temperature T. InSb provides a big Hall constant R_H of about 400 cm³/As for 20°C, but it decreases to 50 cm³/As for 100°C. This fact reasons to assemble two components as a differential transducer. Though this measure produces impressive improvements still further electronic efforts need to be taken to cope with temperature effects. A circuit with two field plates in a bridge and an amplifier with temperature dependent gain is shown in Fig. 16.1.2-7. Increasing temperature T causes the unloaded output voltage U_0 to drop but the internal resistance R_i of the bridge also decreases. The quotient $i = U_0/R_i$ however, stays fairly constant. The remaining dependence on temperature may be encountered by designing

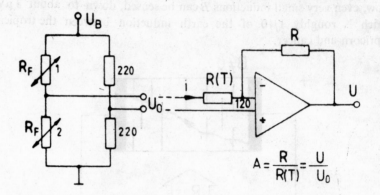

Fig. 16.1.2-7.   Temperature Compensated Circuit with Field Plates R_F, SIEMENS Type FP 212 L 100 (Differential Type).

the gain A of the following amplifier as being temperature dependent, too, with the help of $R(T)$, to ensure compensation.

16.2 SENSING OF FIELD STRENGTH

So far the induction B has been considered. It is normally the quantity of interest, because this quantity is basically needed for most applications of magnetic fields. But the field strength H is usually considered to be its cause and is in need to be measured, too. There are some transducers available to measure H, making use of different effects, being described hereafter.

16.2.1 Saturation Core Transducer (Magnetometer, Flux Gate)

Saturation Core Transducers are known as magnetometers or flux gates. They do not fundamentally depend on temperature, and their features are not subject to tolerances of their components as a matter of first concern. Consequently they provide access to measurements of very weak fields H_x as low as 10^{-4} A/m. Compared to the natural field strength on earth of about 30 A/m, the sensitivity of saturation core transducers is impressive.

Fig. 16.2.1-1 shows the design of the sensor. An iron core 1 of very high permeability is positioned inside an excitation coil 2 which is surrounded by an induction coil 3. The device is most sensitive for an aligned core to the field H_x to be measured. The excitation coil 2 may be fed by a triangular current i. If no field is present ($H_x = 0$), the magneto motive force of coil 2 excites the core symmetrically into positive and negative saturation (continuous line). The related time graph $\phi(t)$ derives identical shapes for both half waves within one cycle. The induction coil 3 responds to the flux changes which induce a related voltage u. For this case see the continuous line of u.

Subjecting the sensor to the field strength H_x of interest causes the core to be premagnetized. The field strength H_x is superimposed on the exciting field of coil 2. This shifts the magnetizing curve $\phi(i)$ to the right, if $H_x < 0$, see Fig. 16.2.1-1, graph of $\phi(i)$, dashed line. (But it would be found to the

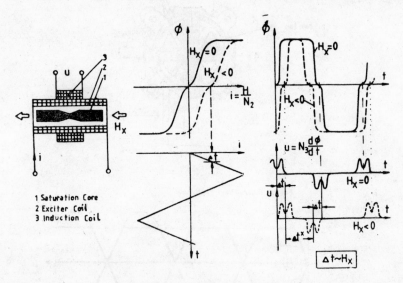

1 Saturation Core
2 Exciter Coil
3 Induction Coil

Fig. 16.2.1-1. Saturation Core Transducer, Configuration of Design and Graphs of Quantities.

left for $H_x > 0$, not depicted.) For the case of $H_x < 0$, see graph, positive saturation is reached only for a short spell of time, but negative saturation continues comparatively longer. Time instants of fast rate of change for the flux are shifted compared to the related instants of time for the case $H_x = 0$. This shift can be sensed with coil 3 delivering u-pulses which are shifted with respect of time by Δt.

The exciting current i was chosen to change with a constant rate to obtain a measurement signal Δt which is proportional to the measurand: $H_x \sim \Delta t$. The shift of the magnetizing curve is translated directly into a time shift.

Looking back at Fig. 16.2.1-1 the neck of the core remains to be reasoned. The narrow cross-section area towards the centre ensures partial saturation which spreads once the field strength increases. This causes an inclined point of inflection for the magnetization curve $\phi(i)$ for small field strength which may be useful to obtain well defined voltage pulses $u(t)$.

The voltage traces $u(t)$ may be evaluated either by investigating the time shift Δt between the two curves, as suggested here, or one could use the last trace only and evaluate for the relative displacement Δt^x of two subsequent positive and negative half waves. —Another method of investigating

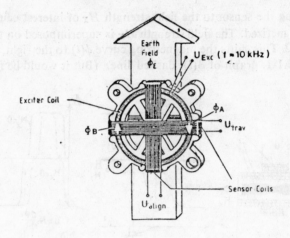

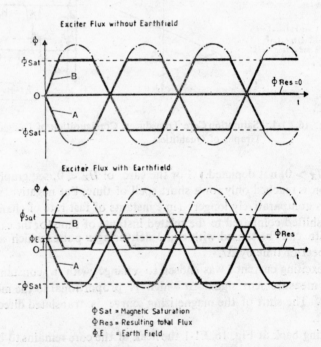

Fig. 16.2.2-1.   FOERSTER Probe as Compass (VDO Adolf Schindling AG, Germany).

the amplitude of the second harmonic of u as depending on H_x is in use, too.

16.2.2 FOERSTER Probe Sensing Field Direction

The FOERSTER Probe is an especially designed saturation core transducer with two induction coils mounted perpendicularly to each other on the toroidal excitation core of high permeability. Being horizontally used, the probe may sense the direction of the earth field, thus serving as a compass. The sensitivity is high enough to investigate a field strength as low as 1/1000 of the natural field of earth, providing about 30 A/m.

The excitation coil is fed from a sinusoidal voltage. It generates a flux which saturates the core in both directions, ϕ_S and $-\phi_S$, as shown in Fig. 16.2.2-1. This is symmetrical if no outer field is present (upper graph). The flux ϕ_B equals $-\phi_A$ at any instant of time. So the resulting flux ϕ_{res} passing through the two induction coils is zero and no voltage appears across them.

But if the field is applied, say in the direction of north (as depicted), $+\phi_B$ is supported to saturate the core earlier during the positive half wave. However, the negative flux $-\phi_B$ needs more time to reach the saturation point. For the opposite flux ϕ_A this is the other way round, respectively. Now a total flux results (ϕ_{res}) being different from zero. It will generate a voltage U_{trav} within the traversely mounted induction coil. The other coil being in line with the field cannot induce a voltage ($U_{alig} = 0$) because the symmetry of its fields was not disturbed by the field of interest. Turning the sensor to the east direction, the relations would be the same for the respective coils then. For any other direction α of the probe the induction coils provide the components of the field vector. A signal processing unit, such as a microprocessor, may calculate the field direction from the component signals which are shown in Fig. 16.2.2-2 as they depend on

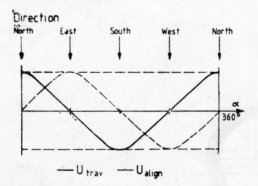

Fig. 16.2.2-2. Component Signals U_{trav} and U_{align} of FOERSTERS onde Depending on the Angular Course α towards Magnetic North,

the course (α). Fig. 16.2.2-3 shows the block diagram of the electronic

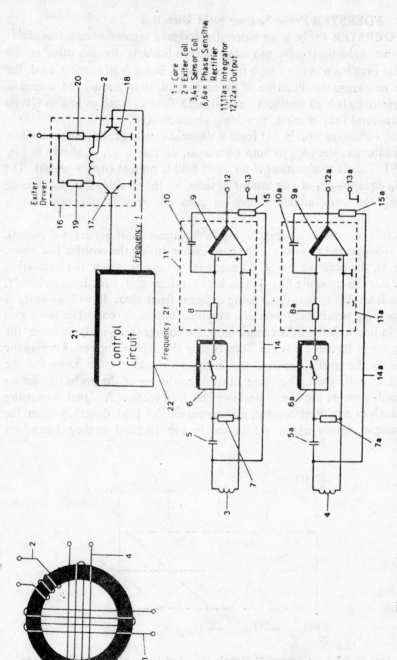

Fig. 16.2.2-3.   Electronic Circuit of Foerster Sonde (VDO Adolf Schindling AG, Germany)

1 = Core
2 = Exiter Coil
3,4 = Sensor Coil
6,6a = Phase Sensitive Rectifier
11,11a = Integrator
12,12a = Output

circuit. The field of the exciter coil 2 is normally superimposed by the field of interest, which causes an unbalance of the flux inside core 1 with a certain direction. The signal voltages of the induction coils 3 and 4 are rectified with phase sensitive rectifiers 6 and 6a. As the signals 3 and 4 provide double the frequency of the exciter, see Fig. 16.2.2-1, the rectifiers work with doubled speed. Their output signals are integrated (11, 11a). A current is derived from the integrator outputs (12, 12a) which is fed back into the induction coils (3, 4), to compensate the measured flux at the very region of core 1 around which coils 3 and 4 are wound. This measure automatically provides for the same working point of the core, whether a field to be measured is present or not and therefore non linearities of the core material cannot effect the measurement. At the same time the integrator delivers a voltage, which is proportional to the related component of the field.

16.2.3 Resistivity of Amorphe Ferroresistive Metal Layers

Latest technology allowed to design a new sensor providing a good resistance change for an applied strength of magnetic field to be investigated. A silicon substrate is used to deposit on its surface a pattern of wide permalloy strips (80% Ni, 20% Fe) being separated by narrow metal lines of good electric conductivity, see hatched areas of Fig. 16.2.3-1. The metal between the permalloy strips forces the current to take the shortest distance through the permalloy with poor electric conductivity. If no field-strength is present the direction of the current will be $+45°$ towards the main strip dimension for strips 1 and 4, see dashed arrows. For strip 2 the configuration was chosen to have the current inclined by $-45°$ and for strip 3, too.

Applying a field only in x-direction turns all current arrows towards this direction, s. dash-dotted lines. By means of an outer yoke guiding the flux a certain state of the field can be arranged to have all dash-dotted currents of the same amount, as depicted. All elements 1 to 4 provide the same resistance in this case. The output voltage U_0 of the bridge will be zero therefore.

As soon as the bridge sensor is additionally subjected to the unknown field strength being applied in y-direction, the current arrows turn towards the resulting field direction then, see unbroken lines. For elements 1 and 4 the current paths get shorter this way, but they are longer for elements 2 and 3. From this fact a decreased resistance for elements 1 and 4 results and an increased resistance for 2 and 3. This produces an output voltage U_0 directly depending on the measurand which is the field strength in y-direction.

Ex. A certain ferroresistive sensor, VALVO type KMZ 10 B, provides a sensitivity of $S = 4 \text{ mV}/(\text{V} \cdot \text{kA/m})$. It is fairly independent of temperature. Fed with constant current, the offset voltage across the output may change by -0.1% per 1 K temperature change. The rated temperature ranges between

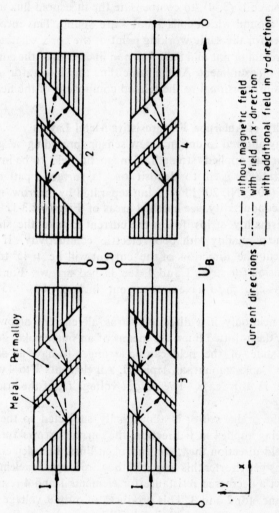

Fig. 16.2.3-1.   Bridge with Ferroresistive Sensor Strips in Barberpole configuration.

— 25°C and +125°C. Linearity is given as 0.5% at 50% full scale range being reached for $H_y = 3$ kA/m for a chosen $H_x = 3$kA/m.

16.2.4 WIEGAND Probe (Bistable Magnetic Sensor)

The Wiegand probe makes use of a straight fine piece of Vicalloy wire (10% V, 52% Co, 38% Fe) which has undergone a special mechanical treatment to produce a bistable behavior of its magnetic features. A slim coil of 1500 turns, and a total length of 15 mm may be wound around the Vicalloy 'core' of about 0.3 mm diameter. This device, when subjected to a changing field, may produce sudden voltage peaks, as shown in Fig. 16.2.4-1. The state of magnetism is changed instantaneously for all constituents of

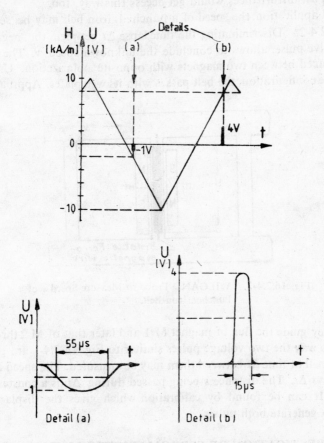

Fig. 16.2.4-1. WIEGAND Probe producing sudden voltage peaks resulting from a bistable magnetic behaviour.

the whole lattice structure. As soon as the first crystal cell collapses, to give way to the outer field it produces an avalanche effect for all other cells—(all in line)—which change their polarization. This happens in nearly no time.

The resulting fast change of flux generates impressive voltages:—1 V within 55 μs for the negative direction and about 4 V or more within 15 μs for the positive direction. The time integrals of both voltage pulses are equal. But the longer pulse time of the negative voltage indicates a slower response to the outer field than for the positive pulse. The lattice structure is easily ready to perform a fast in line reaction for a positive field. However, it is somewhat slower for a negative field.

The Wiegand effect is independent of the rate of change of the applied field $H(t)$. A sensor making use of this effect is superior to other inductive sensors because it may even be used down to zero speed. However, the very fast pulses require a wide frequency range of the processing electronic circuit. Fast disturbances would get access this way, too.

In an application, the speed of a punched iron belt may be sensed, see Fig. 16.2.4-2. Discriminating the time lapse Δt between the negative and the positive pulse, allows to conclude the unknown speed v. The Wiegand coil is placed between two magnets with opposite polarization. Underneath this probe configuration, the belt passes with its web plates. Approaching the

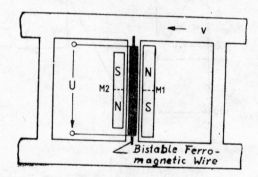

Fig. 16.2.4-2.   WIEGAND Probe to Measure Speed v of a Punched Iron Belt.

sensor they guide the flux of magnet M 1 and later that of M 2 through the coil. This way the two voltage pulses similar to Fig. 16.2.4-1 are generated and the spell of time Δt between them may be measured. The speed is obtained as $v = s/\Delta t$. The distance s being passed during Δt, is a constant of this system. It can be found by calibration which gives the displacement(s) needed to generate both pulses.

16.3   INVESTIGATION OF FERROMAGNETIC MATERIALS

The features of ferromagnetic materials depend considerably on the way we investigate them. Therefore standardized methods have been established. They are quite a few in number and it may easily be found that each application makes use of its own procedure of investigation. But one common statement can be given: The ideal apparatus to measure ferromagnetic constants of different materials should be designed in a way as to allow the magnetic quantities to be set up in the most natural way. A circular structure of the magnetic path serves this purpose. A core shaped as a toroid would not force the field lines to go any other way than that of its own design, which prevents any strays. Therefore, the behaviour of such a core is completely in line with theory.

However, from a practical point of view, toroidal cores cannot be used in all cases. Probes are preferably shaped as straight rectangular cores, thus introducing problems which can only be tolerated as long as similar apparatus is used to repeat the investigation at another place to verify the quantity of interest. If reliable measurements are needed, their limitations should be carefully observed. —The following presentations of ferromagnetic investigations are a limited selection of many more methods. Here we confine ourselves to a few basic instruments.

16.3.1   Hysteresis Loop on the Screen of an Oscilloscope

The function of the flux density B, as it depends on the magnetic field strength H, is called the Hysteresis Loop of the iron core under investigation. It is typical for certain materials and provides the necessary data for designing a magnetic path. The hysteresis loop can be depicted using an oscilloscope. Being independent of time, the trace is a Lissajous figure. A voltage U_y, which is proportional to the flux density B [Vs/m²], is applied to the y-plates and a voltage U_x, proportional to H [A/m], is connected to the x-plates, see Fig. 16.3.1-1. The area of the loop gives the hysteresis losses

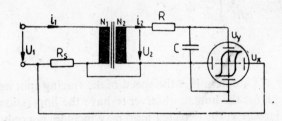

Fig. 16.3.1-1.   Hysteresis Loop of an Iron Core on the Screen of a
Scope.

W_H [Ws/m³] for one magnetic cycle per 1 m³ of iron material. The equation of the units shows this:

$$[W_H] = [B] \cdot [H] = 1 \frac{Vs}{m^2} \cdot 1 \frac{A}{m} = 1 \frac{Ws}{m^3}$$

Also the remanence induction B_r, as intersection point of the positive y-axis, and the coercitive force H_c, as intersection point of the negative x-axis, are available.

The following derivation shows how the voltages U_y and U_x are obtainable using the above circuit. The core under test carries primary (N_1) and secondary (N_2) windings. The current i_1 magnetizes the core along its length l of the magnetic path and causes a voltage drop U_x across the standard resistor R_s. U_x is proportional to H and is used therefore to supply the x-plates as

$$U_x = i_1 R_s \qquad\qquad Hl = i_1 N_1$$

$$= \frac{R_s l}{N_1} H \qquad\qquad i_1 = \frac{H}{N_1} l$$

The secondary windings N_2 produce a voltage U_2 according to the induction law which is proportional to the change of the flux. But instead of the change of flux, the flux, itself is needed. So an integrating network consisting of R and C is employed to obtain a voltage U_y which is proportional to ϕ and to B as well.

$$U_Y = \frac{1}{C} \int i_2 \, dt + C_1 \quad \text{and} \quad i_2 = \frac{U_2}{R} \quad \text{if} \quad \frac{1}{\omega C} \ll R$$

$$= \frac{1}{CR} \int U_2 \, dt + C_1$$

But $\qquad U_2 = N_2 \dfrac{d\phi}{dt}$

So $\qquad U_Y = \dfrac{N_2}{RC} \displaystyle\int \dfrac{d\phi}{dt} \, dt + C_1$

$$= \frac{N_2}{RC} \phi + C_1$$

The frequency of U_1 determines the speed of the tracing spot on the screen. It is fast enough for the human observer to have the impression of a static presentation. However, the depicted loop may include, not only the hysteresis losses of interest, but also considerable eddy current losses, depending on the type of probe and on the frequency of investigation.

The quasi-static methods of chapters 16.3.2 and 16.3.3 are free of this problem. But as soon as the AC line is involved to feed the device of investigation, as for chapter 16.3.4, and the previous paragraph, the question arises how to separate hysteresis and eddy current losses. This is dealt with in chapter 16.3.5.

16.3.2 Curve of Commutation Obtained from Ballistic Flux Meter
The action principle of the ballistic galvanometer was presented in chapter 16.1.1. It is used to investigate the hysteresis loop (curve of commutation) of an iron core (Fig. 16.3.2-1). Prior to measurements the core should be

carefully demagnetized. The bipolar switch S allows to do this by switching it over several times. At the same time the primary magnetizing current I is gradually reduced to zero by help of resistor R_H.

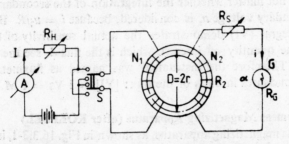

Fig. 16.3.2-1.   Measurement of Curve of Commutation using Ballistic Galvanometer (Flux Meter).

Now, starting with zero, current I may be increased stepwise. For each new step the current I may be reversed with switch S allowing to obtain negative and positive flux $-\phi$ and $+\phi$. The change of flux $\Delta\phi$ results as 2ϕ, generating the secondary voltage which causes the current i through the total loss resistance $R = R_2 + R_S + R_G$:

$$u_2 = N_2\frac{d\phi}{dt} = iR$$

$$\underbrace{\int_{-\phi}^{+\phi} d\phi}_{\Delta\phi = 2\phi} = \underbrace{\frac{R}{N_1}\int_{0}^{t_i} i\,dt}_{Q_B = C_B \cdot \alpha}$$

$$\phi = \frac{RC_B}{2N_2}\,\alpha$$

Side Equations:

with $i = u_2/R$

$$\underbrace{\int_{-\phi}^{+\phi} d\phi}_{} = \underbrace{\frac{1}{N_2}\int_{0}^{t_i} u_2\,dt}_{[VI] = 1Vs}$$

The induction $B = \phi/A$ is caused primarily by the related magnetizing current I which corresponds with the field strength H inside the circular core. H can be calculated using **BIOT-SAVART**'s law for the magnetic circulation:

$$\oint H\,ds = IN_1$$

$$H\int_{0}^{2\pi} r\,d\varphi = IN_1$$

$$H = \frac{N_1}{2\pi r}\,I$$

$2r$ equals the mean diameter of the core.

Both magnetic quantities B and H of the iron probe are available from their related readings α and I, and the graph $B(H)$ may be presented to show their specific dependence on each other, this way characterizing the probe.

It does not matter whether the integration of the secondary current i or of the secondary voltage u_2 is considered, because $i = u_2/R$. But taking the voltage integral (VI) demonstrates the actual sensitivity of the apparatus towards the quantity of interest, which is the flux ϕ, see the previous side equations. Therefore the instrument was named as fluxmeter and consequently it measures in units of the flux: $[VI] = 1 \text{ Vs} = 1 \text{ M}$.

16.3.3   Siemens Magnetizing Apparatus (after KOEPSEL)

The Siemens magnetizing apparatus, as shown in Fig. 16.3.3-1, is based on the moving coil instrument. But now the current (I_c) through the moving coil is constant and the induction B_a of the air gap changes with the current I which allows to magnetize the probe bar. Inside the iron bar the induction B of interest is set up being proportional to B_a which is actually sensed. Due to the yoke assembly measuring errors may arise producing an air gap induction B_a which is not fully proportional to the induction B inside the probe. The compensation coils CC are arranged to compensate this effect. They are series connected to the magnetizing coil MC of opposite winding sense. Ba is therefore quite weak and non-linearities of the yoke ($B(H)$) may only play a minor role. The reading α is proportional to B. — The current I excites the field-strength H of the probe bar. The related reading for I gives H.

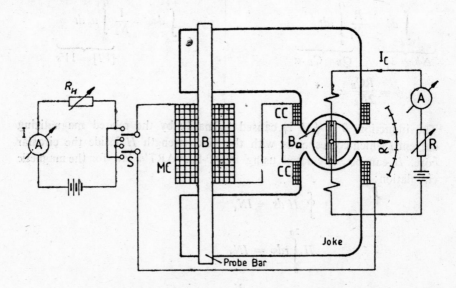

Fig. 16.3.3-1.   SIEMENS Magnetizing Apparatus (after KOEPSEL).

16.3.4 Direct Measurement of Iron Losses

The previously described quasi-static methods to investigate the iron losses of a certain core may be replaced by a device working dynamically, see Fig. 16.3.4-1. It is fed with an AC sine generator G and therefore allows to make use of ordinary volt and watt meters. Both coils should cover the whole core. They may provide the same number of turns $N_1 = N_2 = N$.

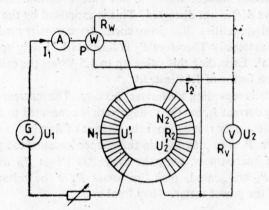

Fig. 16.3.4-1. Direct Measuring of Iron
Losses (Dynamic Method).

The secondary side is loaded by the internal resistances of the voltmeter (R_V) and the voltage path of the wattmeter (R_W) being parallel connected: $R_P = R_V \cdot R_W/(R_V + R_W)$. The voltage $U_2'(t)$ is induced inside the secondary coil giving its RMS value U_2' resulting from the sinusoidal flux $\phi = \hat{\phi} \sin \omega t$ as follows:

$$U_2'(t) = N \frac{d\phi}{dt} = N\hat{\phi}\omega \cdot \cos \omega t$$

$$U_2' = \frac{\omega}{\sqrt{2}} N\hat{\phi} = \underbrace{\frac{2\pi}{\sqrt{2}}}_{4.44 \,=\, 4F_\sim} fN\hat{\phi}$$

However, U_2 can be sensed only:

$$U_2' = U_2 + I_2 R \qquad \text{with} \quad R = R_2 + R_P$$

$$= U_2 \left(1 + \frac{R}{R_P}\right) \qquad \text{with} \quad R_P = \frac{U_2}{I_2}$$

Equating both relations for U_2' gives the maximum induction $\hat{B}$ of interest (with the cross section A of the core):

$$4F_\sim fN\hat{\phi} = U_2\left(1 + \frac{R}{R_P}\right) \quad \text{with} \quad \hat{B} = \frac{\phi}{A}$$

$$\hat{B} = \frac{1 + \dfrac{R}{R_P}}{4F_\sim fNA} U_2$$

$\hat{B}$ is one information needed. It should not exceed certain limits to make sure that the flux $\phi(t)$ is not distorted. This is expressed by the form factor $F_\sim$ of sinusoidal quantities. But iron cores are normally made from non-linear magnetic materials. Therefore $F_\sim$ is taken as 1.11 only up to an induction of 0.8 Vs/m². Exceeding this value up to 1.5 Vs/m² the calculations are performed with a factor of 1.14 instead.

We need details regarding wattmeter reading. The current path carries the magnetizing current I_1, but the voltage path is connected to the secondary voltage U_2. As U_2 only results from the change of flux inside the core, the indicated power P does not contain the copper losses of the primary coil. They are not of interest now. Only the hysteresis losses P_H and the eddy current losses P_E are sensed, plus the losses P_2 of the voltmeter and the voltage path of the power meter. They total as

$$P = P_I + P_2 = P_I + \frac{U_2^2}{R_P} \quad \text{with} \quad P_I = P_H + P_E$$

The iron losses P_I are obtained from the wattmeter reading P and the one of the voltmeter U_2 as

$$P_I = P - \frac{U_2^2}{R_P} \quad \text{(for } N_1 = N_2)$$

For a certain mass m of the iron core the losses per kg of weight are normally given as P_I/m.

In case the number of turns of primary and secondary turns differ from each other, the power needs to be transformed to the primary side as if a turns ratio of $N_1/N_2 = 1$ is valid. Instead of the indicated power P the corrected value $P' = P \cdot N_1/N_2$ needs to be taken into account then.

16.3.5 Separation of Hysteresis and Eddy Current Losses

Most methods to inquire for material constants of ferromagnetics make use of the induction law. Therefore a change of flux needs to be present which is sensed. The iron to be investigated is assembled as a core. Being a conductive material it allows eddy currents which may cause losses W_E due to the resistivity of the material. They are proportional to the squared frequency f^2 of the supply and the squared maximum induction $\hat{B}^2$. —The area of the hysteresis loop A_H [Ws/m³] of the material gives the losses W_H due to the periodic commutation of the magnetic state. They depend on frequency f

directly. The bigger the core volume V the higher are the losses. All losses total as

$$W = W_H + W_E = k_H V A_H f + k_E V \hat{B}^2 f^2$$
$$= C_H f + C_E f^2$$

The squared dependence becomes a linear one by dividing both sides of the equation by f.

$$\frac{W}{f} = C_H + C_E f$$

To obtain the straight line of this function the total losses $(W/f)_1$ and $(W/f)_2$ should be measured at two different frequencies, say at $f_1 = 40$ Hz and $f_2 = 60$ Hz.

The graph of W/f as function of frequency f, (Fig. 16.3.5-1), can be drawn. The elongation of the straight line through the ordinate axis provides the

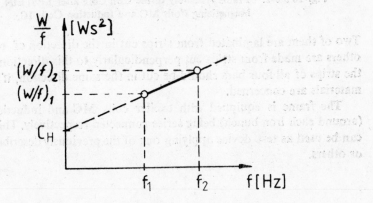

Fig. 16.3.5-1. Separation of Hysteresis and Eddy Current Losses.

constant c_H which allows to calculate the hysteresis losses for the frequency of interest, e.g. for 1 Hz. The area of the hysteresis loop will be rounded once in one second in this case. For the line frequency the hysteresis losses are 50 times higher.

To perform this procedure a generator G of variable frequency is needed. A set of machines consisting of an AC generator driven by a DC motor may easily be controlled to provide the necessary change of frequency.

16.3.6 Core Assembly as a Frame (EPSTEIN Configuration)

During the process of manufacturing magnetic materials as sheets or strips, continuous measurements of magnetic material constants are needed. For fast access of measurement data an easy fitment of the core to be investigated is desirable. A simple assembly together with the exciter and induction coils supports this. A configuration which fits most needs sufficiently was proposed by Epstein. It is an accepted standard now and makes use of four bars of iron probes being assembled as a frame, (Fig. 16.3.6-1). All four bunches of strips measure (30×500) mm having a weight of 2.5 kg each.

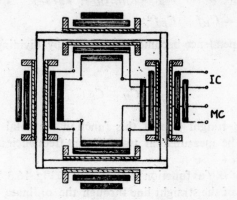

Fig. 16.3.6-1.   Frame Assembly of the Iron Core after EPSTEIN with Magnetizing Coils MC and Induction Coils IC.

Two of them are laminated from strips cut in the direction of rolling. The others are made from strips cut perpendicularly to this direction. However, the strips of all four bars should be cut in the same direction, if cold rolled materials are concerned.

The frame is equipped with exciter coils MC and induction coils IC (around each iron bunch) being series connected respectively. This assembly can be used as test device applying one of the previously described methods or others.

17 Measurements of Selected Non-Electrical Quantities

Measurements of non-electrical quantities are manifold. In fact they are far-reaching and may easily cover a book of their own. So a course of electrical measurements can only deal with some non-electrical quantities and can just point out a few action principles. The ones which are presented here were selected due to their commercial importance and due to the fact that they combine basic measurement ideas. It is felt that the student should know them because they may enable him to find other approaches on his own, even for those quantities which are not dealt with in this chapter.

17.1 GENERAL APPROACH

The non-electrical quantity is usually picked up with the help of a primary sensing element, effecting the actual transducer, which transforms it into an electrical quantity. A pressure, for instance, may be picked up from a diaphragm which bends. An applied strain gauge will undergo a resistance change thus providing a quantity which is measurable by making use of electrical means. Once the problem of transferring the quantity, to be determined, into an electrical one is solved the technique of electrical measurements may be applied which was dealt with in all previous chapters. This chapter deals mainly with the two stages of sensing and transforming. The latter may be performed either actively (the transducer generates energy) or passively (the transducer controls energy which is supplied by a voltage or a current source).

In order to convert a physical quantity into an electrical one, a wide variety of transducer methods may be employed which have quite often equal merits. But each manufacturer has usually specialized on a particular method. A certain choice is often favoured because the instrument under suggestion may be able to reject external disturbing noise, or because it can be battery powered or because it is resistant against aggressive media, etc. Anyway, many reasons need to be considered to choose the right instrument. The development engineer should suggest all this in advance to proceed in the right direction. It may easily happen that he has to undergo a troublesome process of gaining experience, before he can aim for a proper solution.

17.2   FORCE MEASUREMENTS USING STRAIN GAUGES

17.2.1   Gauge Design and Action Principle

Strain gauge grids are usually etched from a thin metal layer on a carrier foil. The grid thickness ranges typically between 8 to 15 μm. For connection purposes the gauge filament has wide terminal areas to facilitate soldering of heavier lead wires to the fine grid. The carrier material is usually nitrocellulose-impregnated paper, but for slightly higher temperature applications, an epoxy backing might be used. Strain gauges as such are applied to the object under investigation using a nitrocellulose solvent release type cement for paper backed gauges or other adhesives recommended by the gauge manufacturer. The application area might be straight or even bent, which is one of the major superiorities of strain gauges over other force measurement devices.

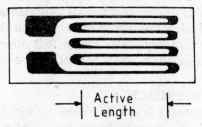

$$\overset{\mid}{\underset{\text{Length}}{\text{Active}}}\overset{\mid}{}$$

Fig. 17.2.1-1.   Etched Strain Gauge on Carrier Foil.

After having fixed the strain gauge properly, it sticks firmly to the object under investigation and is subject to the same relative displacement $\Delta l/l$ if there is any.

Usually a strain or a force determination needs to be executed. It is effected by resistance changes ΔR of the strain gauge due to the elongation or compression of the gauge grid, mainly of its active length. ΔR is usually measured as a voltage U_0 employing a bridge. Thus the gauge is basically an electromechanic transducer.

17.2.2   Resistance Change due to the Applied Force

The nominal value of the gauge resistance R is determined by the specific resistivity ρ of the grid metal, the total length of the filament l and its cross section area A_s

$$R = \frac{l}{A_s}\rho.$$

All variables may cause a resistance change

$$\Delta R = \frac{\partial R}{\partial l}\Delta l + \frac{\partial R}{\partial A_s}\Delta A_s + \frac{\partial R}{\partial \rho}\Delta\rho$$

$$\Delta R = \frac{\rho}{A_s}\,\Delta l - \frac{\rho l}{A_s^2}\,\Delta A_s + \frac{l}{A_s}\,\Delta\rho$$

$$\frac{\Delta R}{R} = \frac{\Delta l}{l} - \frac{\Delta A_s}{A_s} + \frac{\Delta\rho}{\rho} \tag{1}$$

Applying a length change Δl means also to cause a change of the cross section area ΔA_s and a change of the specific resistance $\Delta\rho$. ΔA_s and $\Delta\rho$ depend on Δl, which is dealt with in the following section. For simplicity reasons a circular filament cross-section area of the diameter d is assumed.

Elongation $\Delta l/l$ Causes

Travers Contraction $\Delta A_s/A_s$	*Specific Resistance Change $\Delta\rho/\rho$*

$$A_s = \frac{\pi}{4}\,d^2$$

$$\frac{\Delta A_s}{\Delta d} = 2\,\frac{\pi}{4}\,d$$

$$\frac{\Delta A_s}{A_s} = \frac{2\,\frac{\pi}{4}\,d\Delta d}{\frac{\pi}{4}\,d^2}$$

$$= 2\,\frac{\Delta d}{d}$$

Under tension $(+\Delta l)d$ is diminished by Poisson's ratio ν and consequently A_s, too:

$$\nu = -\frac{\Delta d/d}{\Delta l/l}$$

$$\curvearrowright \frac{\Delta A_s}{A_s} = -2\nu\,\frac{\Delta l}{l}$$

Δl and Δd change the volume V of the gauge wire also. This causes specific resistance changes $\Delta\rho$, described by the Bridgeman ratio C due to inter-crystal displacements inside the metal latice.

$$C = \frac{\Delta\rho/\rho}{\Delta V/V}$$

$$V = \frac{\pi}{4}\,d^2 l$$

$$\frac{\Delta V}{V} = 2\,\frac{\Delta d}{d} + \frac{\Delta l}{l}$$

$$= -2\nu\,\frac{\Delta l}{l} + \frac{\Delta l}{l}$$

$$\curvearrowright \frac{\Delta\rho}{\rho} = C(1-2\nu)\,\frac{\Delta l}{l}$$

Inserting these results into equation (1) gives

$$\curvearrowright \frac{\Delta R}{R} = \frac{\Delta l}{l} + 2\nu\,\frac{\Delta l}{l} + C(1-2\nu)\,\frac{\Delta l}{l}$$

$$= \underbrace{(1 + 2\nu + C - 2C\nu)}_{K} \frac{\Delta l}{l}$$

For constantan $\nu = 0.3$ and $C = 1.13$ making $K = 2.052$. For all grid metals used K comes up very close to 2, so

$$\frac{\Delta R/R}{\Delta l/l} = 2 \text{ may be used generally for } K.$$

The so called K-factor gives the relative resistance change with respect to the relative displacement (elongation).

The relative resistance change $\Delta R/R$ is directly proportional to the elongation $\varepsilon = \Delta l/l$ of the material to which the gauge is bonded. HOOK's law allows to calculate the strain $\sigma = E \cdot \varepsilon$ from ε or $\Delta R/R$ respectively, as well as the force F which is applied to the device under test, once its cross section area A is known. E means the YOUNG's modulus of the device under test. For ordinary steel $E \approx 210\,000 \text{ N/mm}^2$.

$$\frac{\Delta R}{R} = K \frac{\Delta l}{l} = K \cdot \varepsilon$$

$$= K \frac{\sigma}{E}$$

$$= \frac{K}{E} \cdot \frac{F}{A}$$

Fortunately ΔR may easily be measured employing simple bridge circuitry. This fact makes the strain gauge measurement method quite simple and the equipment needed quite economical, which provide some superiorities over other methods.

17.2.3 Full Bridge Circuit

The full bridge circuit is most commonly used. It eliminates disturbing temperature effects for the force measurements. In combination with a certain application pattern the strain gauges permit to suppress certain components of force if need arises. The following chapter gives an example for both.

A lever, as shown in Fig. 17.2.3-1, should be used to measure only the vertical component F_v of a force F. The horizontal component F_H is to be suppressed. Temperature changes should not effect the measurement. The output voltage U_0 should be as high as possible but should be zero if no force is applied.

A full bridge circuit employing four active strain gauges $R_1..., R_4$, serves the purpose. R_1 and R_4 are applied to the upper side of the lever, R_2 and R_3 to the lower side. They are interconnected through the bridge. The gauges R_1 and R_4 are strained whereas R_2 and R_3 are stressed in opposite

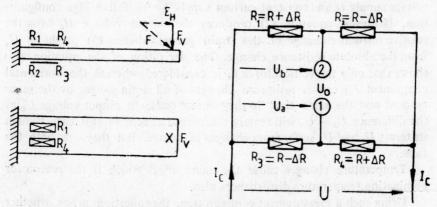

Fig. 17.2.3-1. Force Measurement Device for Determination of the Vertical Component. Temperature Effects Eliminated.

direction resulting in resistance changes $+\Delta R$ and $-\Delta R$ respectively, due to the vertical force component F_v to be measured. The output voltage U_0 is obtained from U_b and U_a. The difference $(U_b - U_a)$ equals U_0. U_0 is zero for zero-force because all strain gauges show equal resistances as their nominal values R. As soon as a force is applied, R_1 and R_4 increase by $+\Delta R$ whereas R_2 and R_3 decrease by $-\Delta R$, resulting in a

$$U_0 = U_b - U_a \neq 0$$

The bridge is supplied by a constant current I_C which produces the voltage U. So U_b and U_a may be expressed by employing U or I_C, equation (2) and (3).

$$U_b = U \frac{R + \Delta R}{2R} \qquad U_a = U \frac{R - \Delta R}{2R}$$

Putting these equations into U_0 gives

$$U_0 = U \left(\frac{R + \Delta R}{2R} - \frac{R - \Delta R}{2R} \right)$$

$$U_0 = U \frac{\Delta R}{R} \qquad (2)$$

The total resistance of the circuit remains constant at the value of R as all resistance changes compensate each other in the upper and in the lower branches. So $U = I_C \cdot R$ producing

$$U_0 = I_C R \frac{\Delta R}{R}$$

$$U_0 = I_C \Delta R \qquad (3)$$

Equations (2) and (3) indicate no substantial difference between constant voltage supply U and constant current supply I_C for full bridge configuration. However, equation (2) produces the output voltage U_0 from the relative resistance change of the strain gauge, whereas (3) produces U_0 from the absolute resistance change. The derivation of the equation also shows that only the vertical force F_V is considered whereas the horizontal component F_H causes resistance changes of all strain gauges by the same amount and the same sign. So they cannot cause an output voltage U_0 as the difference $U_b - U_a$ will remain unchanged at zero in this case though its terms U_b and U_a suffer from changes of course. But they are both the same.

Temperature changes cause the same effect which is the reason for eliminating temperature disturbances also.

Using such a measurement configuration, the question arises whether the nominal resistance R and the change ΔR of a single strain gauge might be determinable from the readily assembled circuit without dismantling anything. The reply is positive as shown here, using a resistance measurement instrument. Having connected it, e.g. across R_1 the resulting R_{res} for zero force and its change ΔR_{res} for an applied force may be measured. Parallel to R_1 the series circuit of R_4, R_3 and R_2 is connected. That means

$$R_{res} + \Delta R_{res} = R_1 // (R_4 + R_3 + R_2)$$

$$= \frac{(R + \Delta R)(3R - \Delta R)}{4R}$$

$$= \frac{3R^2 - R(\Delta R) + 3R(\Delta R) - (\Delta R)^2}{4R}$$

$$= \frac{3R^2 + 2R(\Delta R)}{4R} \qquad (\Delta R)^2 \text{ neglected}$$

$$R_{res} + \Delta R_{res} = \frac{3}{4}R + \frac{1}{2}\Delta R \qquad\qquad (4)$$

The comparison of both sides of this equation (4) gives the expected relations

$$R = \frac{4}{3} R_{res}$$

$$\Delta R = 2\Delta R_{res}$$

These results can also be used for an already assembled force transducer, in order to determine its pin configuration, in case the data sheet is not at hand.

We have shown the principle of rejecting temperature effects. But due to tolerances concerning R of the gauges, temperature influences are eliminated up to a certain degree only. For getting better results quite often an AC supply is used instead of the DC supply so far dealt with. Then there is no need for equal R of the gauges anymore, but for equal ΔR only. Other effects such as disturbing thermo electric voltages generated somewhere in the bridge are also effectively eliminated as they are usually constant.

Mainly supply frequencies of 5 kHz or even 20 kHz are in use. They limit the measurement frequency, i.e. of vibrations, to quite low values. So the advantage of rejecting DC-disturbances by using an AC supply, competes on one hand against the disadvantage of accepting a low bandwidth of the measurand only and on the other hand quite complex circuitry is employed, especially a phase sensitive rectifier which is needed for distinguishing positive and negative load (see also chapter 17.3.3).

DC supplied bridges however may easily. tackle 200 KHz vibrations This high limiting frequency can be obtained from quite a simple circuitry which may be purchased at a reasonable price. For a vibration analysis usually no DC component needs to be considered anyhow. So the drift of a DC amplifier can be allowed as long as the amplifier stays in its linear range of operation.

17.2.4 Half Bridge and Quarter Bridge
There are so called half bridge transducers and quarter bridge transducers available. In these cases there are two or one active strain gauges necessary respectively. The other resistor components are not represented by strain gauges. They may simply be fixed resistors inside the housing of the measurement instrument. Following the same calculation procedure as in chapter 17.2.3, the sensitivity will result only in half of a full bridge or in a quarter of it even, respectively. See also chapter 17.2.6. ($U_0 = U \cdot \Delta R/2R$ for a half bridge configuration and $U_0 = U \cdot \Delta R/4R$ for a quarter bridge circuit). This disadvantage pays off especially in cases which are only in need of a limited accuracy. The fewer number of strain gauges and the simpler connection cable (with three or even two wires only) between the transducer and the indication and supply unit is cheaper, of course, than the accessories for a full bridge configuration.

As there are good reasons for the use of only one strain gauge, occasionally the manufacturers offer them as different types for application on different metals. If they are meant for the use on a test object made from steel, only such elements should be applied which undergo the same expansion as steel does, being subject to temperature increases. This ensures that relative movements between the gauge and the device under test are omitted, thus avoiding fictitious relative displacements ε which are not caused by the force on the tension to be measured.

A question arises for quarter bridges whether a voltage or a current supply should be used. One can prove that the voltage supply is far superior. The linearity is considerably better than that for a current supplied gauge. Certainly an output signal change is caused by the resistance change of the strain gauge in any case, but the transfer factor which describes the transformation from elongation into resistance change is not constant, thus causing a non-linear behaviour. This is far worse for a current supplied quarter bridge than for a voltage supplied one. The latter provides better stability of its sensitivity especially for heavy preloads of the strain gauge.

17.2.5   Practical Side Notes for the Use of Strain Gauge

The amount of $\Delta R/R$ changes should be mentioned. To make sure the repeatability of each measurement, the grid elements should undergo an elongation within their proportionality range to achieve linearity between ΔR and F and to avoid hysteresis. This may be guaranteed for $\Delta R/R$ changes up to 0.05% of R. But as the tolerances of R may easily exceed 0.1%, the measurement effect would possibly be hidden within these tolerances. To avoid this, at least one balanced potentiometer needs to be added within the circuit. It eliminates R tolerances and sets the output voltage U_0 to zero for the unloaded transducer, see also Fig. 7.3.5-3.

The fact that ΔR changes are normally small makes it necessary to ensure the stability of strain gauge features and their connections. This can be achieved by careful application of the gauges to the device under test in combination with a proper connection of the cable electrically as well as mechanically. Usually manufacturers of strain gauges give application and fixing procedures which should be strictly followed. In order to avoid side effects from mechanical loads, which the connection cable might carry, it may be quite feasible to use fixed soldering points that serve as interconnections between the actual gauge and the cable. To avoid corrosion the whole connection and mounting should finally be covered using silicon rubber, for instance.

Last but not least, we should mention the wide variety of strain gauge applications. Torques may be measured as well as pressures or other mechanical quantities. In these cases special construction means are employed to convert the measurand into a displacement Δl, which is the actual quantity, the gauge responds to.

The relative displacement $\Delta l/l$ is normally extremely small. So strain gauges are usually considered as motionless transducers, though there is a slight motion necessary. To have nearly no motion is usually a demand for achieving a high life time of the transducer.

17.2.6   Example of a Force Transducer
(Using two active Strain Gauges and two Dummies)

The construction of the transducer becomes self-evident from Fig. 17.2.6-2. The load to be measured is applied to the lower hole of 4.5 mm diameter. The strain gauges 1 and 4 are electrically connected to the dummies 2 and 3 (which are also applied to the transducer) and to the supply voltage U as Fig. 17.2.6-1 shows.

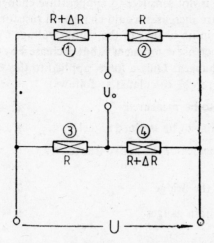

Fig. 17.2.6-1.   Bridge Circuit referring to the Force
Transducer of Fig. 17.2.6-2.

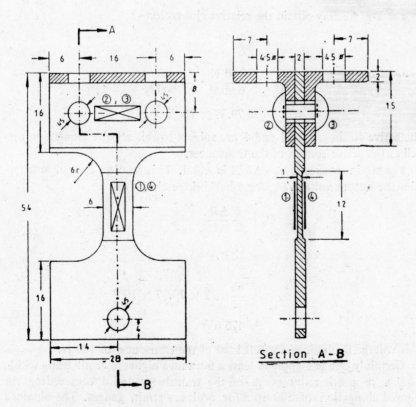

Section A-B

Fig. 17.2.6-2.   Strain Gauge Transducer (Housing Omitted).

The measurement is not sensitive to temperature changes and to bending loads: Temperature increases would change all resistors 1 to 4 by the same value, resulting in no change of the output voltage U_0. Bending loads would, for instance, increase component 1, but decrease 4 by the same amount thus preventing U_0-changes. Only a force, applied to the transducer causes an output voltage U_0, to be calculated as follows:

Maximum Force to be measured $\qquad\qquad$ $F = 100$ N

Transducer Material to be strained: Steel $\qquad$ $E = 210\ 000$ N/mm²

Cross section measures of the active transducer
part $\qquad\qquad\qquad\qquad\qquad\qquad\qquad\qquad$ $A = (6 \times 1)$ mm²

Supply voltage of the bridge $\qquad\qquad\qquad$ $U = 6$ V

K-Factor of the strain gauges $\qquad\qquad\quad$ $K = 2$

Tension $\quad \sigma = \dfrac{F}{A}$

$\qquad\qquad = \dfrac{100\ \text{N}}{6\text{mm}^2}$

As $\sigma = \varepsilon \cdot E$ we may obtain the relative elongation

$$\varepsilon = \frac{\sigma}{E}$$

$$= \frac{100\ \text{N}}{6\ \text{mm}^2} \cdot \frac{\text{mm}^2}{210000\ \text{N}}$$

$$= 79.37 \times 10^{-6}.$$

The active strain gauges 1 and 4 are subject to this relative elongation ε as well as the active section of the transducer.

For strain gauges $K \cdot \varepsilon = \Delta R / R$ is valid. This relation is used to determine the output voltage U_0 for a half bridge circuit as

$$U_0 = \frac{U}{2} \frac{\Delta R}{R}$$

$$= \frac{U}{2} K \cdot \varepsilon$$

$$= \frac{6\text{V}}{2} \times 2 \times 79.37 \times 10^{-6}$$

$$\simeq 476\ \mu\text{V}$$

This voltage is obtained for full load of the transducer of 100 N.

Certainly one can apply at least a five times higher load still being within Hook's proportionality range of the transducer and not exceeding the allowed elongation of $500 \cdot 10^{-6}$ for ordinary strain gauges. The obtained output voltage U_0 would then be five times higher. But even 2.5 mV is still a low voltage. One could increase it by a factor of two employing a full bridge circuit if one proceeds only for a high output voltage. But a DC amplifier would still be necessary. It can be purchased at a reasonable price providing a gain of up to 200 which would be sufficient to drive robust equipment either for indicating the force to be measured with the help of a voltmeter or for further processing of the data obtained.

17.3 DISPLACEMENT MEASUREMENT USING DIFFERENTIAL COIL SENSOR

17.3.1 Transducer Fundamentals and its Design

The inductivity L of a coil, having an iron core, depends on the length l of the magnetic flux-lines (ϕ) and other variables of course, as shown in Fig. 17.3.1-1.

The current I generates a magnetic circulation V_m which in turn causes a magnetic flux ϕ. Both quantities determine the magnetic resistance:

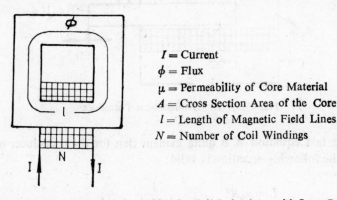

I = Current
ϕ = Flux
μ = Permeability of Core Material
A = Cross Section Area of the Core
l = Length of Magnetic Field Lines
N = Number of Coil Windings

Fig. 17.3.1-1. Principal Assembly of a Coil Inductivity with Iron Core.

$$R_m = \frac{V_m}{\phi} \qquad V_m = I \cdot N \qquad \phi = B \cdot A$$
$$= \mu H \cdot A$$
$$= \frac{I \cdot N}{\mu \dfrac{IN}{l} A} \qquad\qquad = \mu \frac{IN}{l} A$$

$$R_m = \frac{l}{\mu A}$$

This expression for R_m gives the inductivity, which is defined by the following equation

$$L = \frac{N^2}{R_m} \qquad\qquad \mu = \mu_0 \mu_r$$

$$= N^2 \mu_0 \mu_r \frac{A}{l}$$

$$L = N^2\mu_0 \frac{1}{\dfrac{l}{\mu_r A}}$$

The formula shows the possibility to achieve an L-change by changing the field length l, a fact which will be used for the following derivation. A device doing this is called a displacement sensor employing a (differential) coil transducer which provides a displaceable armature, e.g. a core inside a coil.

A transducer permitting practical displacement measurements l_L is depicted in Fig. 17.3.1-2.

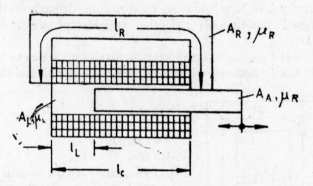

Fig. 17.3.1-2.   Displacement Transducer.

Using the last equation it is quite evident that for the transducer of Fig. 17.3.1-2 the following equation is valid:

$$L = \mu_0 N^2 \frac{1}{\dfrac{l_L}{\mu_L A_L} + \dfrac{l_C - l_L}{\mu_A A_A} + \dfrac{l_R}{\mu_R A_R}}$$

For smallest possible air gap length ($l_L = 0$), L becomes its maximum value L_{max} which will be used as a reference.

$$L_{max} = \mu_0 N^2 \frac{1}{\dfrac{l_C}{\mu_A A_A} + \dfrac{l_R}{\mu_R A_R}}$$

$$\frac{L}{L_{max}} = \frac{1}{1 + \dfrac{l_L}{l_C} \dfrac{\dfrac{l_C}{\mu_L A_L} - \dfrac{l_C}{\mu_A A_A}}{\dfrac{l_C}{\mu_A A_A} + \dfrac{l_R}{\mu_R A_R}}}$$

Employing the same material for all iron parts ($\mu_R = \mu_A = \mu_{Fe}$) and making all cross section areas equal ($A_R = A_A = A_L$) and using a construction that provides $l_C \approx l_R \approx l_{Fe,\,max}/2$ (which is better for smaller l_L) and knowing that $\mu_{Fe} \gg \mu_L$ we obtain from the last equation

$$\frac{L}{L_{max}} = \frac{1}{1 + \dfrac{l_L/\mu_L}{L_{Fe, \ max}/\mu_{Fe}}}$$

This formula is formally equivalent to $y = 1/(1 + ax)$ with $y \sim L$ and $x \sim l_L$, depicted in Fig. 17.3.1-3(a).

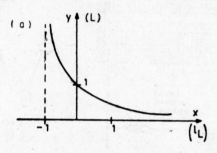

(a)   **Single Coil Type**

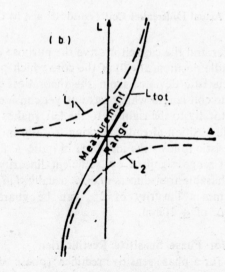

(b)   **Differential Coil Type.**

Fig. 17.3.1-3.   Characteristics of Coil Transducers.

A considerable non-linear characteristic is evident for this type of a single coil transducer. Using two coils, circuited as series connection the difference of two inductivities will be measurable producing a remarkable linearization on one hand and even a higher sensitivity on the other. The slope of the total characteristic L_{tot} is much steeper [Fig. 17.3.1-3(b)], which is basically valid for an actual transducer, as shown in Fig. 17.3.1-4.

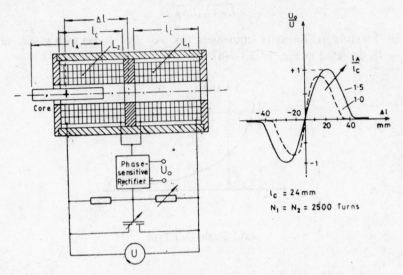

Fig. 17.3.1-4. Actual Differential Coil Transducer and its Characteristic.

The potentiometer and the capacitor serve the purpose of balancing to make $U_o=0$ for middle position($\Delta l=0$) of the core which produces $L_1=L_2$.

The real characteristic deviates from the theoretical one for high Δl. This is due to a finite coil length which was not yet considered. Of course, taking out the core totally to the right or to the left, makes L_1 equal to L_2 too (as for its middle position already), making $U_o = 0$ in this case, too. So the real characteristic follows the ideal one in quite a limited range of Δl, the measurement range, signified by sufficient linearity depending on the ratio l_A/l_C. Purchasable transducers realize usually $l_A/l_C = 1.5$. For a coil length l_C of 24 mm a linearity of 2% can be guaranteed within a measurement range Δl of $\pm$ 10mm.

17.3.2. The Need for Phase Sensitive Rectification

To explain the need for a phase sensitive rectifier (phase discriminator) a quarter strain gauge bridge may serve as an example. A DC-supply current I_C feeds the bridge. When applying a positive force F_1 to the transducer a change of ΔR_1 causes positive output voltage U_{01}, and for negative F_2 a negative U_{02} is available. The amounts of U_{01} and U_{02} reflect directly the magnitudes of F_1 and F_2. The signs of U_{01} and $- U_{02}$ carry the information about the direction of the forces F_1 and F_2.

For AC-supply current $I_\sim$ a $U_{0\sim}$ is obtained. Its amplitude contains the information about the magnitude of the force. This is for sure. But what happens to a force of opposite sign? Again a polarity change is effected, see

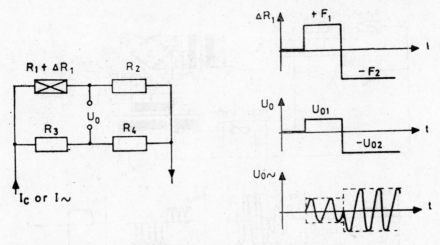

Fig. 17.3.2-1. Quarter Strain Gauge Bridge and Output Voltage U_0 for DC-and AC-Current Supply.

$U_{0\sim}$ which has a phase shift of 180° exactly. So the changed phase relation of $U_{0\sim}$ obviously carries the information about the direction of force.

17.3.3 Phase discriminator

The use of ordinary rectifiers would unfortunately cause the loss of the important sign information, as $|U_{0\sim}|$ for positive and negative force looks quite the same. In order to avoid this a phase sensitive rectifier is usually employed. A circuit used quite often, is given in Fig. 17.3.3-1, the so called phase discriminator.

Again the time graphs are shown for sudden ΔR_1 changes as in the previous chapter. The quarter strain gauge bridge and the bridge rectifier B are fed from the transformer T_1 which is supplied by a voltage U of sufficiently high frequency f_H. The auxiliary voltage U_a across the bridge B switches on the upper two valves of B and the lower two valves consecutively. The amplified gauge voltage $A \cdot U_{0\sim}$ is of the same frequency as U_0. $A \cdot U_{0\sim}$ equals $U_B \cdot U_B/2$ is obtainable from the two secondary windings of T_2. Its middle terminal is connected directly to the low pass LP (lower lead). The potential of the upper T_2-terminal is switched to the upper lead of the LP for conductive upper valves of B during the first half wave of U_a. Its second one switches on the lower valves of B, thus transmitting the potential of the lower T_2-terminal to the upper lead of LP as well. This way a full bridge rectification of U_B is obtained, see U_{LP}. For positive ΔR_1 all half waves are positive and for negative ΔR_1 they are all negative. So the original information is maintained. The low pass smoothes its input voltage U_{LP} and produces the output voltage U_0.

The phase shifter should also be mentioned. The gauge supply voltage and the auxiliary voltage U_a are processed by different elements following

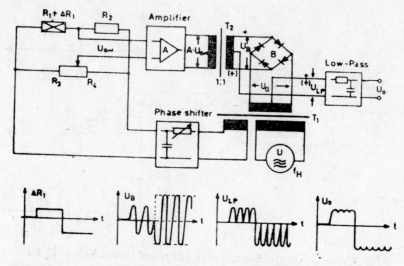

Fig. 17.3.3-1.   Phase Discriminator and its Time Graphs of different Voltages.

different information paths. This may cause phase shifts between these voltages, which would result in phase-cut half waves of U_{LP} producing a reduced U_0. To achieve maximum possible U_0 the phase shifter allows to set U_0 to high values by avoiding phase cutting. This ensures the full sensitivity of the circuit.

We chose an AC-powered strain gauge bridge as an example for the use of a phase discriminator. Of course the same circuit may be employed for differential coil transducers to sense their inductivities being subject of the core position, see Fig. 17.3.1-4.

17.4   VELOCITY MEASUREMENT WITH INDUCTION COIL TRANSDUCER

Direct velocity measurement of a linear movement is usually effected employing the induction law. Either the motion of a permanent magnet relative to a fixed coil is sensed or vice versa. The first type is presented in Fig. 17.4-2 (a) showing the simplest possible electromagnetic transducer and the second in (b) showing the electrodynamic transducer principle.

The permanent magnet (N-S) generates a constant flux ϕ. But (due to the relative movement) its change $d\phi/dt$ generates a voltage u across the windings N of the coil proportional to the velocity of the movement, see Fig. 17.4-1 which is drawn for $N = 1$. These types of transducers are simple and therefore quite robust. Especially the type (a) allows quite a lot of windings which ensure a sufficient high output voltage u. It is actively generated. So there is no need for a supply. As the coil is fixed its connection leads do not suffer any motion.

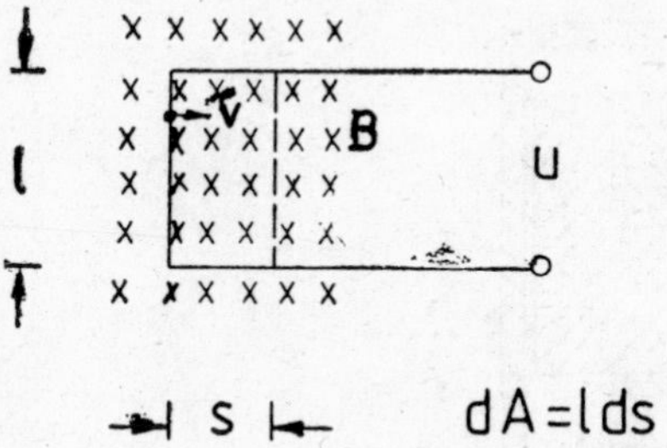

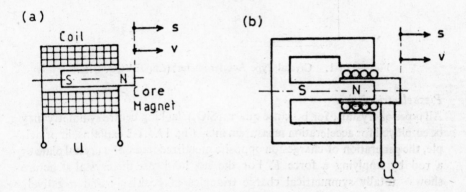

$$U = -N\frac{d\phi}{dt} = -NB\frac{ds}{dt}$$

$$= -NBl\frac{ds}{dt} = -NBl \cdot v$$

$$= C \cdot v$$

Fig. 17.4-1.   Voltage Generation of a loop in a Field B.

(a) Coil

S — N Core Magnet

(b)

S — N

Fig. 17.4-2.   Velocity Transducer Principles.

If instead of the velocity, the displacement is needed an integrator needs to be employed additionally, see chapter 17.6. In fact its output will be proportional to the displacement but this is true for alternating displacements only. A static one cannot be sensed for there is no velocity. In this case the velocity sensor would produce zero output independently of the magnet position and the integrator output would indefinitely drift away.

17.5   ACCELERATION MEASUREMENTS

17.5.1   Crystal Transducers
Crystal Transducers are normally used in dynamic measurements from approximately 2 Hz to 1 MHz. They generate a high output voltage but with an extremely high internal source impedance. Normally, they are used in combination with a charge amplifier, as discussed later. They are robust and small (smallest diameter 5 mm). An actual transducer is shown in Fig. 17.5.1-1. It engages a heavy preload with the help of a nut and a rod, this way ensuring one polarity of the output voltage even for very high alternating accelerations. An inert mass m loads the crystal C, making the device   o the acceleration to be measured.

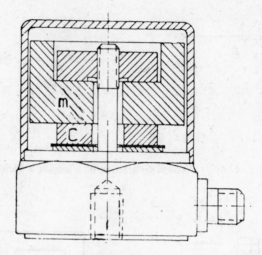

Fig. 17.5.1-1.   Crystal Type Accelerometer. (ENDEVCO, U.K.)

Piezoelectric Effect

All types of crystals, for instance quartz (SiO_2), lacking centre symmetry may be employed for acceleration measurements. Fig. 17.5.1-2 explains, in principle, the generation of charges on opposite metalized faces of a crystal plate or a rod by applying a force F. For the no load case the crystal structure shows totally symmetrical charge triangles of positively and negatively charged ions. Their centres are found at the same location. As they are of opposite sign to each other the crystal appears to be neutral from outside.

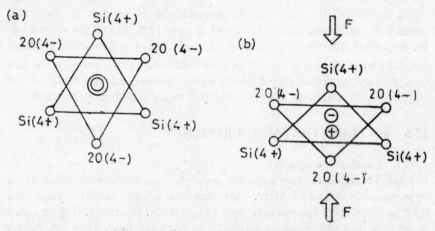

Fig. 17.5.1-2.   Quartz-Lattice-Structure, unloaded (a) and loaded by Force F (b).

Loading the crystal by an outer force F causes a displacement of the ions. The previously equi-sided charge triangles undergo a deformation and become equi-leged ones. This causes a displacement of the positive and the negative charge centres. The metalized outside faces get charged and an electrostatic voltage across them may be measured. (Conversely, the application of an electric field may cause the crystal to contract or to expand, depending on the polarity of the field).

Charge Amplifier

In order to measure the charge generated by a crystal-transducer usually a circuit, as shown in Fig. 17.5.1-3, known as charge amplifier, is employed. It makes use of an operational amplifier having a high open-loop-gain A. The following derivation shows the proportionality of the output voltage

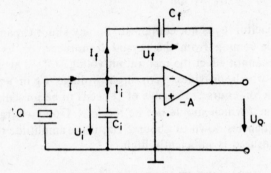

Fig. 17.5.1-3.   Charge Amplifier.

U_0 to the input charge Q. Changes of Q produce the input current I which is the sum of the two capacitor currents I_f and I_i. There is no current going into the amplifier element, as was already shown in chapter 10.

$$\frac{dQ}{dt} = I = I_f + I_i$$

The feed back current I_f is due to the voltage change across the feed back capacitor C_f: $I_f = C_f dU_f/dt$. But of course $U_f = U_i - U_0$ and $U_0 = -A \cdot U_i$, so

$$I_f = C_f \frac{dU_f}{dt} = C_f \frac{d(U_i - U_0)}{dt} = -C_f \frac{\left(\frac{U_0}{A} + U_0\right)}{dt}$$

$$I_f = -C_f \left(1 + \frac{1}{A}\right) \frac{dU_0}{dt}$$

The current I_i is due to voltage changes across the input capacitor C_i which may again be expressed by the output voltage $U_0 = -A \cdot U_i$.

$$I_i = C_i \frac{dU_i}{dt}$$

$$I_i = -\frac{C_i}{A} \frac{dU_0}{dt}$$

Substituting both currents I_f and I_i of the first equation produces

$$\frac{dQ}{dt} = -\left[C_f \left(1 + \frac{1}{A} \right) + \frac{C_i}{A} \right] \frac{dU_0}{dt}$$

$$U_0 = -\frac{Q}{C_f \left(1 + \frac{1}{A} \right) + \frac{C_i}{A}}$$

An operational amplifier usually provides nearly infinite gain A, so the real output voltage U_0^* is obtained as

$$U_0^* = \lim_{A \to \infty} U_0 = -\frac{Q}{C_f}$$

The input capacitor C_i is not contained. So any shunt capacity of a long connection cable coming from the crystal transducer, or its shunt conductivity even, cannot effect the real output voltage U^*. This is due to the circuit feature which provides zero input voltage U_i in any case, see chapter 10. But of course, a change of Q needs to be present, generating the input current I which should not be too low. Good charge amplifiers can handle frequencies down to about 2 Hz, if the amplitude of the acceleration to be measured is sufficiently high.

17.5.2 Seismic Differential Diaphragm Transducer

As shown in chapter 2.2 a damped mass spring system may be employed for acceleration measurements. Only the damping D and the mass m need to be small but the spring rate c considerably high. In the case of a small ratio m/c the effected displacement of the mass proves to be directly proportional to the acceleration. These features may easily be realized by employing quite simple means, see Fig. 17.5.2-1.

The diaphragm 4 represents the mass. Punched slots allow to set the mass/spring rate-ratio to a certain value. A diaphragm made from a thin copper-berillium-bronze forms a considerably high spring rate and at the same time a small mass. The internal damping is low as well. Subject to the acceleration the diaphragm bends accordingly. This displacement may be sensed by two coils printed on ceramic plates 2 which are positioned on either side.

The right coil has decreased its inductivity due to eddy currents inside the diaphragm when it sets nearer to it, whereas at the same time the left coils inductivity increases because of the increasing distance towards the diaphragm. Using a phase sensitive rectifier supplied by a 1 MHz-voltage, Fig. 17.5.2-2, the output voltage U_0 will be a direct measure of the applied acceleration. The coils L_1 and L_2 determine the currents I_1 and I_2. Their difference $I_1 - I_2$ charges the output capacitor, resulting in U_0. The mean value of U_0 gives the output information. It may be positive or negative depending on the direction of acceleration. —An actual transducer of the following dimensions showed a considerable sensitivity:

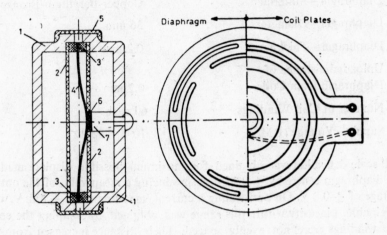

Fig. 17.5.2-1. Diaphragm Accelerometer.

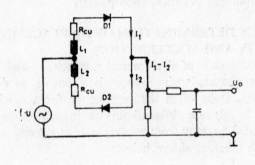

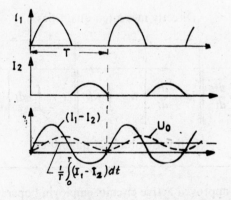

Fig. 17.5.2-2. Phase discriminator for Diaphragm
Acceleration.

Diaphragm—Material	Copper-Beryllium-Bronze
Diaphragm—Diameter	36 mm
Diaphragm—Thickness	0.2 mm
Unloaded distance of Diaphragm from Coils	0.2 mm
Number of Coil Windings	60 each
Supply—Voltage (RMS)	10 V, 1 MHz

Full scale deflection was obtained for maximum possible displacement of the diaphragm centre by ± 0.18 mm, producing a mean value of the output voltage of ± 0.8 V. On performing a current measurement, ± 8 mA were obtainable. Linearity within this range was obtained by shaping the coils. The windings were not evenly spaced. Their distance increased from the edges of the coils towards their centres.

Different ranges of acceleration to be measured are obtainable by choosing the diaphragm thickness appropriately.

17.6 INDIRECT DETERMINATION OF DISPLACEMENT, VELOCITY AND ACCELERATION

The mechanical quantities of displacement s, velocity v, and acceleration a, may be directly measured using appropriate sensors, as described in the previous chapters. Quite often the very sensor for a special measurand is not available. In this case differentiators or integrators may be employed to obtain the needed quantity from another one, as s, v and a have defined relations to each other, as shown below:

Directly measured quantities					
\multicolumn{2}{c}{s}	\multicolumn{2}{c}{v}	\multicolumn{2}{c}{a}
$v = \dfrac{ds}{dt}$	$a = \dfrac{d^2s}{dt^2}$	$a = \dfrac{dv}{dt}$	$a = \int v\,dt$	$v = \int a\,dt$	$s = \iint (a\,dt)\,dt$
\multicolumn{6}{c}{Indirectly obtained quantities}

In Chapter 10.4.4 approved active circuits employing operational amplifiers were shown. They are quite useful for computation of small signals from directly measured quantities. In this case an additional attenuation should be avoided by using active differentiators or integrators. But quite often high sensor output signals are available. They may be processed by employing simple passive *RC* combinations. In order to achieve good performance of computation their output signals will be much smaller than their input signals.

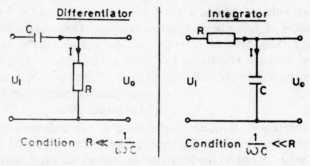

| Differentiator | Integrator |

Condition $R \ll \dfrac{1}{\omega C}$ | Condition $\dfrac{1}{\omega C} \ll R$

The current I is mainly determined by the impedance within the input line:

$$I = C\frac{dU_C}{dt} \qquad\qquad 1 = \frac{U_R}{R}$$

The output voltage U_0 will be considerably small with respect to the input voltage U_i, so

$$U_C \approx U_1 \qquad\qquad U_R \approx U_i$$

$$I \approx C\frac{dU_i}{dt} \qquad\qquad I \approx \frac{U_i}{R}$$

The current I causes the output voltage U_0 which is the voltage drop

Across R \qquad\qquad Across C

$$U_0 = I \cdot R \qquad\qquad U_0 = \frac{1}{C}\int I\, dt + c_1$$

Introducing I produces

$$\boxed{U_0 = CR\frac{dU_i}{dt}} \qquad\qquad \boxed{U_0 = \frac{1}{RC}\int U_i\, dt} + c_1$$

The output is proportional to the differential quotient of the input | The output is proportional to the integral of the input.

Applying a sinusoidal input $U_i = \hat{U}_i \sin \omega t$ produces an output amplitude which depends on the frequency

$$U_0 = \underbrace{\hat{U}_i CR \omega}_{\hat{U}}.\ \cos \omega t$$

$$U_0 = -\underbrace{\frac{\hat{U}_i}{RC\omega}}_{\hat{U}_0}\cos \omega t + c_1$$

The frequency dependance of the output voltage limits the use of these circuits.

Usually disturbing high frequencies are picked up together with the actual signal to be processed. They produce high voltages which might cause saturation of following amplifiers, putting them out of action. This will be avoided effectively by adding a low pass after the differentiator.

The signal to be processed may often be found to be superimposed onto slowly changing offsets, for instance of amplifiers circuited in front of the integrator. They may cause considerable output drifts throughout wide ranges, thus hiding the actual signal of interest. Not employing amplifiers in front of an integrator avoids this effect.

These disadvantages show up all the more if two differentiators or two integrators are used in series. To employ a second circuit might be indicated by the need to obtain the acceleration a from the displacement s or s from a respectively, see table. But having a velocity sensor, only one differentiator or one integrator will be needed to produce either an a-signal or an s signal. This should be stressed especially because a v-sensor can easily be assembled by winding a simple coil, shown in chapter 17.4 and as it is not in need of a supply.

17.7 TEMPERATURE MEASUREMENTS

Temperature measurements have a wide variety of applications. For industrial purposes those are especially of interest which employ thermoelectric transducers. They provide an electrical output signal clearly related to the temperature measured. Such instruments are likewise suitable for indication purposes and for further electrical processing of the obtained signal. Thermocouples, resistance thermometers and thermistors provide useful features to solve innumerable temperature measurement problems.

17.7.1 Thermoelectric Transducers

Thermoelectric transducers consist of two dissimilar metal wires A and B which are welded or brazed together at two junctions thus forming a thermo couple, see Fig. 17.7.1-1.

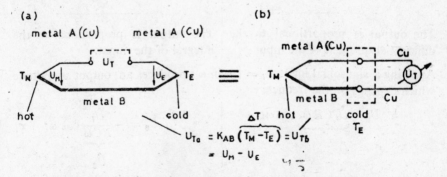

Fig. 17.7.1-1. Thermo Couple (Temperature Difference Measurement)
(a) Principle, (b) Actual Circuit.

SEEBECK-Effect

To one of the junctions the measurement temperature T_M is applied. The other one is subject to the environmental temperature T_E (or a constant reference temperature). Basically there are always two temperatures sensed which effect voltages across the junctions. So actually there are two electrical elements being series connected but opposite to each other. Thus for equal temperatures $T_M = T_E$ no thermo voltage appears: $U_T = 0$. But commonly the temperatures are different and a voltage U_T occurs. It is proportional to the temperature difference.

$$U_T = K_{AB}(T_M - T_E)$$

Calibration of a Thermocouple and Properties of Different Materials

K_{AB} is the thermosensitivity of the couple A/B which may depend on the temperature: $K_{AB} = f(T)$. Due to this fact the last equation can only be applied within a narrow temperature range. Tables are usually needed to derive from the measured voltage the temperature to be determined. The accuracy of this procedure is high. The DIN specification allows an absolute deviation of 3°C from the true value for low temperatures and up to 9°C for very high temperatures. Quite a good survey over applicable temperatures to different thermocouples provides fig. 17.7.1-2 which gives the obtainable voltages, too. But the calibration should always involve the tables of thermoelectric voltages which puts U_T into relation towards T_M for $T_E = 0°C$.

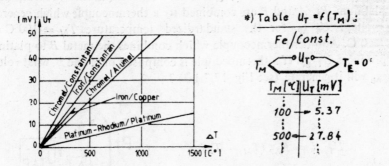

Fig. 17.7.1-2. Thermovoltage over Temperature of Important Material Combinations.

Example for Determination of Measuring Temperature T_M with Table*

Ex. Suppose the reference junction is at $T_E = 100°C$ and the voltage reading is $U_T = 22.47$ mV for an Fe/constantan thermo couple.

$$U_T = K(T_M - T_E) = U_M - U_E$$

$$22.47\text{mV} = K(T_M - 100°\text{C}) \rightarrow \text{into the table: } 100°\text{C} \searrow 5.37\text{mV} = U_E$$

$$= K \cdot T_M - 5.37\text{mV} \qquad\qquad K \cdot T_E = \overbrace{K \cdot 100°\text{C}}$$

$$\searrow K \cdot T_M = (22.47 + 5.37)\text{ mV}$$

$$= 27.84\text{ mV} = U_M \leftarrow \text{out of the table: } 27.84\text{mV} \searrow 500°\text{C} = T_M$$
$$\underbrace{\qquad\qquad}_{K \cdot T_m}$$

$$\searrow T_M = 500°\text{C}.$$

Both table readings could be used to determine the thermosensitivity: $K' = 5.37$ mV/100°C and $K'' = 27.84$ mV/500°C = 5.57 mV/100°C. If the thermo couple was a linear transducer element K' should be equal to K''. As this is not the case temperature measurements are entirely based on empirical calibration tables. U_T depends on the alloy components of the metals, of course. For this reason the tables of the manufactures should only be used.

For automatic temperature determination throughout wide ranges, the formula for U_T which employs the squared thermosensitivity K_2 or even the cubic one K_3 may be useful. But one should be aware that the accuracy using the tables cannot be reached this way.

$$U_T = k_1 \Delta T + K_2 \Delta T^2 + K_3 \Delta T^3 + \dots$$

Pair of Thermocouples

The thermosensitivity is usually given for a certain metal A in comparison to platinum *Pt*. A and P are combined to a thermocouple which generates the voltage $U_{T, AP}$ for the standardized temperatures $T_M = 100°$C and $T_E = 0°$C. Another thermocouple which combines the metal B to platinum Pt produces $U_{T, BP}$. If a thermocouple is composed of A and B, what voltage $U_{T, AB}$ will it generate (see Fig. 17.7.1-3) ?

$$U_{T, AP} = K_{AP} (T_M - T_E)$$

$$U_{T, BP} = K_{BP} (T_M - T_E)$$

$$\underbrace{U_{T, AP} - U_{T, BP}}_{} = \underbrace{(K_{AP} - K_{BP})}_{} (T_M - T_E)$$

$$U_{T, AB} = K_{AB} (T_M - T_E)$$

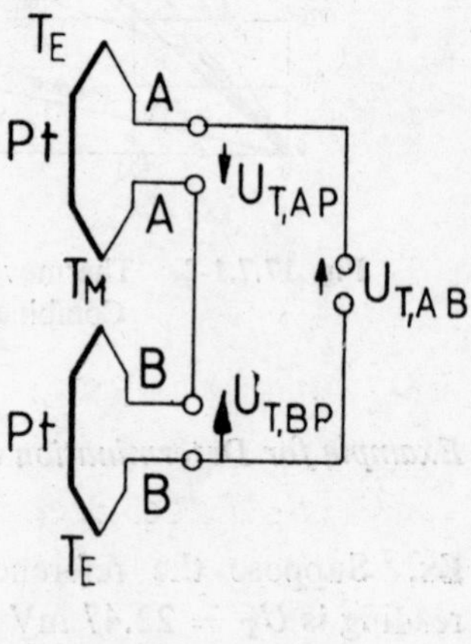

Fig. 17.7.1-3.   Series Connected Thermo Couples Involving a Third Metal (Pt).

This derivation shows that it is sufficient to have the thermosensitivities of different metals towards platinum. If two other metals A and B are combined the sensitivity K_{AB} may be calculated from K_{AP} and K_{BP}.

The third metal (Pt) does not contribute to U_T. So A and B may be connected directly at their T_M and T_E junctions leaving out Pt. Fig. 17.7.1-3 changes into the previous Fig. 17.7.1-1 (a) this way. —Leaving a third metal within the circuit, but making use of copper instead of Pt turns the function of Fig. 17.7.1-3 into that of Fig. 17.7.1-4 (a).

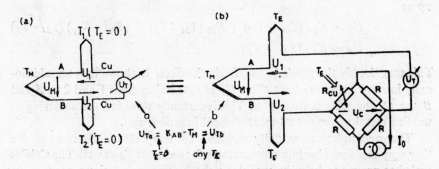

Fig. 17.7.1–4. Thermo Couple Circuits (Absolute temperature Measurements) (a) With Zero Reference, (b) With Compensation Bridge for Environmental Temperature.

Actual Measurement Circuits

The second depiction of Fig. 17.7.1-1 showed that usually the thermocouple A/B is connected to an instrument which involves copper wires. Thus there are actually three joints. Fig. 17.7.1-4 (a) concludes the obtainable output voltage $U_T \cdot T_E$ may be any environmental temperature. The thermovoltage U_T is obtained from $U_T = U_1 + U_2 + U_M$. Assuming that the upper copper wire is connected to A, U_1 may be found as

$$U_1 = K_{\text{CuA}}(T_1 - T) = (K_{\text{CuP}} - K_{\text{AP}})(T_1 - T)$$

Assuming that the lower copper wire is connected to B, U_2 may be found as

$$U_2 = K_{\text{BCu}}(T_2 - T) = (K_{\text{BP}} - K_{\text{CuP}})(T_2 - T).$$

Connecting A and B produces U_M. It can be written quite formally in the same way as U_1 and U_2:

$$U_M = K_{\text{AB}}(T_M - T) = (K_{\text{AP}} - K_{\text{BP}})(T_M - T).$$

T is any temperature of the respective second junction. T needs not to be considered because it does not appear in the final result of U_T.

U_1, U_2, and U_M may be used to find U_T as

$$U_T = K_{CuP}\, T_1 - K_{AP}\, T_1 - K_{CuP}\, T + K_{AP}\, T$$
$$+ K_{BP}\, T_2 - K_{CuP}\, T_2 - K_{BP}\, T + K_{CuP}\, T$$
$$+ K_{AP}\, T_M - K_{BP}\, T_M - K_{AP}\, T + K_{BP}\, T$$
$$U_T = K_{CuP}\,(T_1 - T_2) + K_{BP}\,(T_2 - T_M) + K_{AP}\,(T_M - T_1)$$

In practice the junctions of copper to A and to B are usually both at environmental temperature T_E ($T_1 = T_E$ and $T_2 = T_E$) and so U_T comes up as

$$U_T = K_{BP}\,(T_E - T_M) + K_{AP}\,(T_M - T_E) = (K_{AP} - K_{BP})\,(T_M - T_E)$$
$$U_T = K_{AB}\,(T_M - T_E)$$

This was already the result of the Seebeck-effect. It shows that an additionally inserted metal does not effect U_T in case its junctions 1 and 2 to A and B are at the same temperature. This is certainly valid for the soldering or welding metal of the actual measurement junction, too.

Thermocouples are actually sensitive to temperature differences. Bu they may measure absolute temperatures for constant T_E as well. The Celsius scale has proved to be quite useful. So the reference junctions need to be at 0°C for T_E. For precise measurements of T_M the reference temperature may be provided with the help of a Dewar jar containing a slash of ice water. Admittedly the handling of such a container is not convenient. If the accuracy of the measurement needs not to be high, circuit (b) of fig. 17.7.1-4 provides a very handy instrument. The environmental temperature T_E produces thermoelectric voltages U_1 and U_2. But they are compensated with the help of a bridge which produces U_C. As T_E might change, also U_C needs to be changed. This is effected by a temperature dependent resistance R_{Cu}. The other resistors R are constant. U_T is indicated. It should be proportional to the measured temperature T_M:

$$U_T = \underbrace{U_1 + U_2 + U_M}_{K_{AB}\,(T_M - T_E)} + U_C$$
$$= K_{AB}\, T_M \underbrace{- K_{AB}\, T_E + U_C}_{O\,!}$$

The bridge needs to supply a voltage $U_C = K_{AB} \cdot T_E$.

Pyrometer Thermometer and RMS Current Measurement Device
Thermocouples produce low voltages only. The circuit of fig. 17.7.1-5 multiplies the output quantity by piling up several thermocouples. It is a series connection of several measuring thermocouples and several reference couples (n).

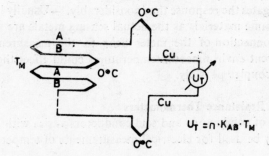

Fig. 17.7.1-5. Thermopile.

If the measurement junctions are assembled at the focal point of an optical system the device can be used to determine the temperature of heated surfaces without physical contact. Such an instrument is a thermal radiation pyrometer.

Another application uses the thermocouple for RMS measurements of HF-currents. For this purpose it is enhoused inside an evacuated enclosure together with a heater which heats it up, once the current to be measured is applied, see fig. 17.7.1-6.

The temperature of the heater is proportional to the RMS value I of the current. So the voltmeter which measures the thermoelectric voltage has a non-linear scale for the indicated I. The spacing between the scale divisions widens for increasing I. The reading is independent of the frequency and therefore also independent of the waveshape. The output is proportional to the current squared.

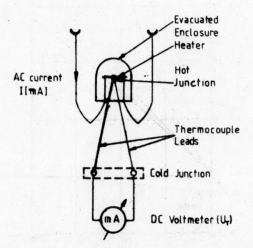

Fig. 17.7.1-6. RMS Current Meter with Vacuum Thermocouple.

Thermocouples may be designed as fast responding elements. But the enclosure elongates the response time considerably. —Usually long extension cables of the same materials as the actual sensing metals are used in order to avoid the connection of the meter near to the measurement location at which different environmental temperatures could effect the connections of the thermocouples possibly.

17.7.2 Metal Resistance Thermometers
The resistance of conductors and semiconductors varies with temperature. This effect may be used for electrical measurements of temperatures.

Temperature Functions of Different Materials
For resistance thermometers different materials are in use, see fig. 17.7.2-1. For metals the resistance usually increases with temperature. Copper, nickel and platinum are most commonly used. But for semiconductor NTC-thermistors, the resistance decreases considerably for increasing temperatures. NTC devices have roughly a ten times higher sensitivity than metal resistance transducers. The latter may be described analytically as

$$R = R_0(1 + \alpha_1 \Delta T + \alpha_2 \Delta T^2 + \ldots + \alpha_n \Delta T^n)$$

$$\Delta T = T - T_0$$

T is the measured temperature and T_0 the reference temperature. R_0 is the resistance at T_0. The linear temperature coefficient for metals is about

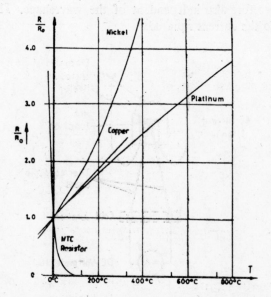

Fig. 17.7.2-1. Temperature Dependence of Resistive Materials.

0.4%/K. Terms of higher order than one are normally neglected if ΔT is small enough.

Platinum is available in pure form. It is very stable and can be used over wide temperature ranges. Specifications are arranged in corresponding standards. A fast response may be reached with a wire wound grid or a free spiral. For most sensors however a glass housing or a metal tube enclosure provides rigidness. But the response is slow then. 10 Ohm⁻, 100 Ohm⁻ and 1000 Ohm⁻ devices are commercially available with an accuracy of 0.01% typically.

Temperature coefficient

Let us assume a linear relation between temperature and resistance, see fig. 17.7.2-2. If the resistance at reference temperature T_0 equals R_0 the resistance $R(T)$ increases by ΔR for a temperature increase of ΔT.

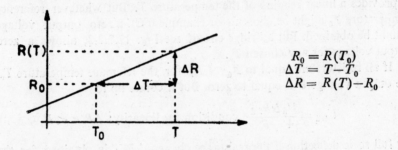

$$R_0 = R(T_0)$$
$$\Delta T = T - T_0$$
$$\Delta R = R(T) - R_0$$

Fig. 17.7.2-2. Linear Resistance Change over Temperature.

$$R(T) = R_0 + \Delta R$$

The slope of the function $m = \Delta R/\Delta T$, which makes

$$R(T) = R_0 + m \cdot \Delta T$$

$$= R_0 \left(1 + \frac{m}{R_0} \Delta T\right) \quad \frac{m}{R_0} = \frac{\Delta R/\Delta T}{R_0}$$

$$= R_0 (1 + \alpha_1 \Delta T) \quad = \frac{\Delta R/R_0}{\Delta T} = \alpha_1$$

α_1 is the linear temperature coefficient of the resistance which gives the relative resistance change per degree temperature change.

Sensitivity

For constant current supply of the thermometer gauge, a voltage drop U appears across the device which changes proportional to the temperature to be measured:

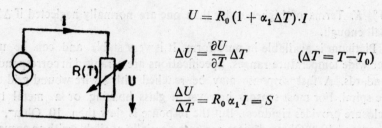

$$U = R_0 (1 + \alpha_1 \Delta T) \cdot I$$

$$\Delta U = \frac{\partial U}{\partial T} \Delta T \qquad (\Delta T = T - T_0)$$

$$\frac{\Delta U}{\Delta T} = R_0 \alpha_1 I = S$$

The sensitivity S of this configuration is defined as voltage change ΔU per degree temperature change ΔT. In order to achieve high sensitivity S, the current I should be high. But self heating needs to be avoided. High R_0 aims also at high S. But the outer dimensions of the device limit R_0. The temperature coefficient α_1 cannot be freely chosen. It is a physical constant.

Thermometer Circuits

One circuit configuration was already shown in the previous paragraph. It provides a linear reading of the temperature T. But whatever reference temperature T_0 might be chosen (for example 0°C), a zero output voltage cannot be obtained. But a bridge circuit, see Fig. 17.7.2-3, allows a zero output voltage for any chosen T_0.

If all resistors are equal to R_0 (of R_T for the reference temperature T_0) the output voltage U is equal to zero. But it comes up to

$$U_{fsd} = \frac{U_0}{4} \frac{\Delta R_{fsd}}{R} \quad \text{condition for linearity: } \Delta R << R$$

for full scale deflection if the resistance change ΔR_{fsd} is obtained for the temperature range ΔT of consideration. The above mentioned formula was derived in chapter 17.2. It is valid for this quarter bridge with only one active transducer element R_T. The supply voltage U_0 should be chosen in such a way as to avoid self-heating of R_T, say for 0.1 W allowed dissipation loss. In case the obtained output voltage U_{fsd} should be considered too low, an amplifier may be employed. The indication may be provided with the help of an ordinary panel meter.

The circuit of Fig. 17.7.2-3 (b) is specially designed for monitoring a certain temperature for which $R_T = R$. In this case the line resistance R_L may even suffer from changes. They cannot effect the reading α of the pointer, see chapter 2.6 for further details.

Series Resistors

Quite often a certain well defined temperature coefficient α of a resistance needs to be realized. For temperature sensing devices α should be high But in electronics, temperature effects usually disturb the performance of. the circuit. In this case α is mostly needed as zero. The realization of a certain α may be effected by combining two elements of different α_a and α_b as series or parallel connection.

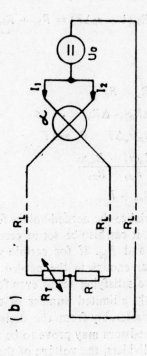

Fig. 17.7.2-3. (a) Bridge-Thermometer and (b) Ratiometer-Thermometer.

$$R_a = R_{a0}(1 + \alpha_a \Delta T) = R_{a0} + \overbrace{R_{a0}\alpha_a\Delta T}^{\Delta R_a}$$

$$R_b = R_{b0}(1 + \alpha_b \Delta T) = R_{b0} + \underbrace{R_{b0}\alpha_b\Delta T}_{\Delta R_b}$$

$$R = R_a + R_b$$

$$\Delta R = \Delta R_a + \Delta R_b$$

$$R_0 \alpha \Delta T = R_{a0}\alpha_a\Delta T + R_{b0}\alpha_b\Delta T$$

$$\alpha = \frac{R_{a0}\alpha_a + R_{b0}\alpha_b}{R_{a0} + R_{b0}}$$

$$R_0 = R_{a0} + R_{b0}$$

The last two equations materialize the needed values for α and R_0. α is not only a function of α_a and α_b, it can also be set to certain values by choosing suitable values for R_{a0} and R_{b0}. If for certain values of R_{a0}, R_{b0}, α_a and α_b the value α is not near enough to the needed one, another choice of R_{a0} and R_{b0} may produce a satisfying result even for maintained values of α_a and α_b. As there is only a limited number of alphas available the results might still deviate considerably from the desired. One in this case a parallel connection of the resistors may prove to be a better choice.

Side note: For voltage dividers, the voltage of the divider point stays constant with temperature if $\alpha_a = \alpha_b$.

Parallel Resistors

$$R = \frac{R_a R_b}{R_a + R_b}$$

$$\frac{\Delta R}{R} = \frac{\Delta R_a}{R_a} + \frac{\Delta R_b}{R_b} - \frac{\Delta R_a + \Delta R_b}{R_a + R_b}$$

$$\frac{R_0\alpha\Delta T}{R_0} = \frac{R_{a0}\alpha_a\Delta T}{R_{a0}} + \frac{R_{b0}\alpha_b\Delta T}{R_{b0}} - \frac{(R_{a0}\alpha_a + R_{b0}\alpha_b)\Delta T}{R_{a0} + R_{b0}}$$

$$\alpha = \alpha_a + \alpha_b - \frac{R_{a0}\alpha_a + R_{b0}\alpha_b}{R_{a0} + R_{b0}}$$

$$= \frac{R_{a0}\alpha_a + R_{b0}\alpha_b + R_{b0}\alpha_a + R_{a0}\alpha_b - R_{a0}\alpha_a - R_{b0}\alpha_b}{R_{a0} + R_{b0}}$$

$$\alpha = \frac{R_{a0}\alpha_b + R_{b0}\alpha_a}{R_{a0} + R_{b0}}$$

$$R_0 = \frac{R_{a0} \cdot R_{b0}}{R_{a0} + R_{b0}}$$

Now in α the resistance of one element is combined with the temperature coefficent of the other. Following the same calculation procedures, as for the series circuit, another temperature coefficient α is available now by using elements of the same α_a and α_b as for the series connected resistors.

RC-Device

The time constant τ of an RC-device needs to provide a certain temperature coefficient α_τ.

$$\tau = R \cdot C$$

$$\frac{\Delta\tau}{\tau} = \frac{\Delta R}{R} + \frac{\Delta C}{C}$$

$$\frac{\tau_0\alpha_\tau\Delta T}{\tau_0} = \frac{R_0\alpha_R\Delta T}{R_0} + \frac{C_0\alpha_C\Delta T}{C_0}$$

$$\alpha_\tau = \alpha_R + \alpha_C$$

In this case only the temperature coefficients of the components determine the total α_τ. But R may be composed of parallel or series components and this choice exists for C also. So usually the demands for a certain temperature behaviour can be met. If, for instance, α_τ should equal zero α_R needs to be negatively equal to α_C.

Commercial R- and C-Components

The following statements of the temperature coefficients of different R- and C-components provide only a rough survey in order to obtain at least an idea of what can be realized. But for exact calculations the manufacturer's catalogues should be referred to.

Temperature Coefficients of Resistors

Wire wound resistors	$(2 \text{ to } 5) \times 10^{-6}/\text{K} = f \text{ (material)}$
Metal film resistors	$(20 \text{ to } 50) \times 10^{-6}/\text{K} = f (T)$
Metal oxyde resistors	$(150 \text{ to } 200) \times 10^{-6}/\text{K} = f \text{ (material, } T)$
Carbon resistors	$-(300 \text{ to } 800) \times 10^{-6}/\text{K} = f (T, R)$

The actual value of the temperature coefficient with the given range may depend on the material of the resistance. Usually, the alphas are not constant but depend on the temperature. Therefore, the resistance does not obey a linear temperature function. This is especially of importance for wide temperature ranges. Carbon resistors provide only negative alphas.

But their values depend also on R. The variations of alpha are manifold. But manufacturer's lists provide reliable data. It should be pointed out that the tolerances of R and α need to be considered, too.

Temperature Coefficient of Capacitors
The variety of available temperature coefficients is quite wide. But for orientation purposes the highest possible positive and negative values should be given and one near zero:

MKT—capacitor	$500 \times 10^{-6}/K$
MKC—capacitor	$70 \times 10^{-6}/K$
Polypropylene foil capacitor	$-300 \times 10^{-6}/K$

17.7.3 Thermistors
Thermistors usually provide a high negative temperature coefficient (NTC). Their resistance may change throughout decades. But within a small temperature interval, a positive temperature coefficient may also be realized (PTC), see Fig. 17.7.3-1. PTCs are preferably used for trigger purposes at certain temperature levels. NTCs are mainly employed for analog measurements through wide temperature ranges.

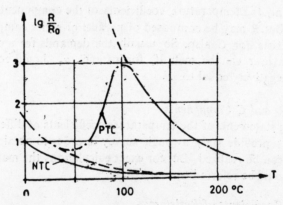

Fig. 17.7.3-1. Temperature Characteristic of an NTC-
and a PTC-Resistance Thermometer.

Features of NTCs
NTCs are semiconductor components made from CuO_2 or other oxides. Their temperature coefficient is about ten times higher than the one of metals but negative. The accuracy of thermistors is not high. 2% may be considered to be good. NTCs cover a temperature range of roughly $-60°C$ to $+150°C$ only. Due to their high sensitivity they are frequently used.

Their time constant may be short as thermistors can be assembled as very small beads. Their long term stability is poor.

Temperature Function of NTC

The resistance of an NTC may be described as

$$R(T) = A \exp(B/T)$$

B is a material constant given in Kelvin, for example $B=3800$ K. A is the resistance for infinite temperature. As this temperature cannot be realized it is quite impractical to use A as a signifying quantity. For this reason another function of $R(T)$ should be developed which meets the practical aspects.

At reference temperature T_0, the NTC produces its nominal value R_0. As only A is unknown it may be determined and one can introduce it into $R(T)$.

$$R(T_0) = R_0 = A \exp(B/T_0) \quad A = R_0 \exp(-B/T_0)$$

$$R(T) = R_0 \exp[B(1/T - 1/T_0)]$$

Ex. Example: For a thermistor the nominal value $R_0 = 230\Omega$ was measured at $T_0 = 20°C = 293$ K. A t $T=21°C=294$ K, the resistance was $R(T)=220\Omega$. Let us determine the material constant B.

$$\frac{R(T)}{R_0} = \exp\left(B\left(\frac{1}{T} - \frac{1}{T_0}\right)\right)$$

$$\ln\frac{R(T)}{R_0} = B\left(\frac{1}{T} - \frac{1}{T_0}\right)$$

$$B = \frac{\ln\dfrac{R(T)}{R_0}}{\dfrac{1}{T} - \dfrac{1}{T_0}}$$

$$= \frac{\ln\dfrac{220\Omega}{230\Omega}}{\dfrac{1}{294\,\text{K}} - \dfrac{1}{293\,\text{K}}} = \frac{-0.04445\,\text{K}}{-1.16087 \times 10^{-5}} = 3829\,\text{K}$$

Temperature Coefficient

The temperature coefficient β_T of an NTC-resistance may also be named as temperature sensitivity. It is obtained from $R(T)$.

$$\beta_T = \frac{\dfrac{\Delta R}{R_T}}{\Delta T} = \frac{\dfrac{dR}{dT}}{R_T} \quad \frac{dR}{dT} = A\left(-\frac{B}{T^2}\right)\exp(B/T)$$

$$= \frac{A\left(-\dfrac{B}{T^2}\right)\exp(B/T)}{A\exp(B/T)}$$

$$= -\frac{B}{T^2} = f(T)$$

As thermistors are non-linear elements the temperature coefficient β_T is a function of the applied temperature, of course.

Ex. Assume an **NTC** with $B = 3829$ K was chosen. Calculate the temperature coefficient β_T for normal temperature of $T_0 = 20°C = 293$ K.

$$\beta_T = -\frac{3829 \text{ K}}{(293 \text{ K})^2} = -4.46 \times 10^{-2}/\text{K} \approx -10\alpha$$

In fact the relative resistance change per degree temperature change is about 10 times higher than the one of a metal type resistance for normal temperature, but negative.

Ex. In case the material constant B is not known β_T may be determined from two measurements of $R(T)$ for two different temperatures. This obtains B first, following the above mentioned procedure. After that β_T will be determined as described. But even a linear approximation might do if the two temperatures are close to each other, $\Delta T = (T - T_0) \to 0$:

$$R(T) = R_0(1 + \beta'_T \Delta T)$$

$$\beta'_T = \frac{\dfrac{R(T)}{R_0} - 1}{T - T_0} \qquad \begin{array}{l} T_0 = 20°C : \quad R_0 = 230\Omega \\[4pt] T = 21°C : R(T) = 220\Omega \end{array}$$

$$= \frac{\dfrac{220\Omega}{230\Omega} - 1}{(294 - 293)\text{ K}} = -4.3 \times 10^{-2}/\text{K} \cdot$$

The comparison of the correct value β_T with this value β_T' of the approximation shows an error of

$$\varepsilon = \frac{\beta'_T - \beta_T}{\beta_T} = \frac{(-4.3 + 4.46) \cdot 10^{-2}}{-4.46 \times 10^{-2}} = -3.6\%$$

Side note: B usually suffers from wide tolerances. 20% are common for non selected types. But $10\%-$, $5\%-$, and even 2% types are available at higher prices.

Linear NTC-Thermometer (hyperbolic approximation)
A linear **NTC**-Thermometer can easily be realized using the circuit of Fig. 17.7.3-2. For this purpose an NTC needs to be chosen providing a temperature function of its resistance R_T which does not deviate from the hyperbola $R = a/T$ considerably. Under this condition the measured current I changes almost linearly with temperature:

$$I = \frac{U_0}{R_T} \qquad R_T = \frac{a}{T}$$

$$= \frac{U_0}{a} \cdot T$$

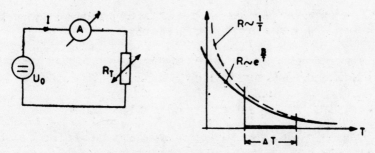

Fig. 17.7.3-2   Quasi-Linear Thermistor Circuit.

As there are innumerable NTCs available, a linear thermometer of this type can be realized for many applications. For certain accuracy requirement the applicable temperature range ΔT is, of course, limited to exactly defined boundaries.

Index